ME

Ba

MEDIA MANUALS

THE ANIMATION STAND
Zoran Perisic

BASIC FILM TECHNIQUE
Ken Daley

BASIC TV TECHNOLOGY
Robert L. Hartwig

CREATING SPECIAL EFFECTS
FOR TV & FILMS
Bernard Wilkie

EFFECTIVE TV PRODUCTION
Gerald Millerson

MOTION PICTURE CAMERA &
LIGHTING EQUIPMENT
David W. Samuelson

MOTION PICTURE CAMERA DATA
David W. Samuelson

MOTION PICTURE CAMERA
TECHNIQUES
David W. Samuelson

MOTION PICTURE FILM
PROCESSING
Dominic Case

PANAFLEX USERS' MANUAL
David W. Samuelson

SCRIPT CONTINUITY & THE
PRODUCTION SECRETARY
Avril Rowlands

16 mm FILM CUTTING
John Burder

SOUND RECORDING
AND REPRODUCTION
Glyn Alkin

SOUND TECHNIQUES FOR
VIDEO & TV
Glyn Alkin

TV LIGHTING METHODS
Gerald Millerson

THE USE OF MICROPHONES
Alec Nisbett

USING VIDEOTAPE
J. F. Robinson & P. H. Beards

VIDEO CAMERA TECHNIQUES
Gerald Millerson

THE VIDEO STUDIO
Alan Bermingham,
Michael Talbot-Smith,
Ken Angold-Stephens & Ed Boyce

YOUR FILM & THE LAB
L. Bernard Happé

Basic
TV
Technology

Robert L. Hartwig

FOCAL PRESS
Boston London

Focal Press is an imprint of Butterworth Publishers.

Library of Congress Cataloging-in-Publication Data

Hartwig, Robert L.
  Basic TV technology / Robert L. Hartwig.
    p. cm.—(Media manuals)
  Bibliography: p.
  ISBN 0-240-80051-6
  1. Television—Handbooks, manuals, etc. I. Title.
  II. Series.
TK6642.H37   1990
621.388—dc20   89–7865

British Library Cataloguing in Publication Data

Hartwig, Robert L.
  Basic TV technology.
  1. Television equipment
  I. Title II. Series
  621.388

  ISBN 0-240-80051-6

Butterworth Publishers
80 Montvale Avenue
Stoneham, MA   02180

10 9 8 7 6 5 4 3 2 1

Printed in the United States of America

Contents

INTRODUCTION
ACKNOWLEDGEMENTS

THE ATOM AND ELECTRICITY 10
  The parts of the atom
  The flow of electrons through
    metals
BASIC CIRCUITS 12
  Direct current (DC)
  Alternating current (AC)
UNITS OF MEASUREMENT (1) 14
  Voltage
  Current
  Power
  Resistance and impedance
  Mathematical symbols and
    formulas
UNITS OF MEASUREMENT (2) 16
  Frequency
  AC frequency
  Capacitance
ABBREVIATIONS 18
  Conversions
INDUCTION AND NOISE 20
  Induction
  Noise
  Signal-to-noise ratio
CAMERA SYSTEM
COMPONENTS 22
  Pickup tubes
  Target
  Cathode
CAMERA SCANNING 24
BLANKING 26
  Horizontal blanking
  Vertical blanking
WAVEFORM DISPLAY 28
CATHODE RAY TUBES (CRTs) 30
  CRT and pickup tube
    relationship
  Need for interlace scanning
COLOR SYSTEM 32
  Color versus black and white
  Additive and subtractive colors
  Complementary colors
  Red, green, and blue tube
    differences
HOW THE EYE SEES LIGHT 34
REGISTRATION AND ENCODING 36
  Registration
  Encoding
  Home video cameras

COLOR CRTs 38
  Convergence
VIDEO SYSTEM COMPONENTS 40
SYNC GENERATORS 42
SYNC GENERATOR SIGNALS 44
  Combining sync with video
VECTORSCOPE 46
  Reading the vectorscope
  Color bar display
SYNC FLOW DIAGRAMS (1) 48
  Distribution amplifiers
SYNC FLOW DIAGRAMS (2) 50
  Termination
  Subcarrier, blanking, and sync
    pulses
CAMERA FLOW DIAGRAM 52
COMBINING SYNC AND
CAMERA FLOW DIAGRAMS 54
  Out of phase cameras
  Timing the system
FILM-TO-VIDEO TRANSFER 56
FILM CHAIN LAYOUT 58
  Condenser lens
  Neutral density filter
FILM CHAIN PROJECTORS AND
VIDEO CAMERAS 60
HIGH-QUALITY FILM-TO-VIDEO
TRANSFER 62
  Flying spot scanner
VIDEO SWITCHERS 64
  Passive switchers
  Vertical interval switchers
  Component switchers
  Special effects
SWITCHER APPLICATIONS 66
  Production and editing
    switchers
  On-air switchers
  Routing switchers
PRODUCTION SWITCHER FLOW
DIAGRAM 68
  Switcher buses
  Switcher outputs
SWITCHER TRANSITIONS AND
SPECIAL EFFECTS 70
  Wipes
SPECIAL EFFECTS KEYS —
LUMINANCE KEYS 72
SPECIAL EFFECTS KEYS —
CHROMA KEYS 74
DIGITAL SPECIAL EFFECTS 76
  Compressions
  Pushes

Flips
Rotations
Other special effects
VIDEOTAPE RECORDING
TECHNOLOGY   78
Recorders
Videotape
Recording heads
VIDEO RECORDING STANDARDS
AND FORMATS (1)   80
Audio versus video recording
Quadraplex video recording
VIDEO RECORDING STANDARDS
AND FORMATS (2)   82
Helical video recording
SMPTE Type C and U-Matic
formats
VIDEO RECORDING STANDARDS
AND FORMATS (3)   84
Other broadcast quality
formats
Sound and control tracks
TIME BASE ERROR   86
EXTERNAL CAUSES OF TIME
BASE ERROR   88
Gyroscopic time base error
TIME BASE ERROR CORRECTION   90
Quad
Helical
VTR LOCKUP (1)   92
Capstan lock
Vertical lock (capstan servo)
VTR LOCKUP (2)   94
Frame lock
Horizontal lock
EDITING VIDEOTAPE   96
Physical cutting and splicing
Electronic editing
THE EDITING PROCESS   98
TYPES OF EDITS   100
Assemble edits
Insert edits
EDITING METHODS   102
Manual editing
Control track counters
SMPTE TIME CODE EDITING   104
OFF-LINE AND ON-LINE EDITING   106
Off-line editing
On-line editing
EDITING BY COMPUTER   108
TIME BASE CORRECTORS   110
Review of time base error
problem
What a time base corrector
does
How a TBC works

DIGITAL TECHNOLOGY — THE
HEART OF A TBC   112
A-to-D conversion
Sampling and quantizing
THE SYNCHRONIZED SIGNAL   114
Horizontal sync as a clock
D-to-A conversion
Video proc amp
Window of correction
LARGER SYNC PROBLEMS AND
SOLUTIONS   116
Nonsynchronous sources
Frame store synchronizer
OTHER ADVANTAGES OF TBCs
AND FSSs   118
Dynamic tracking heads
Freeze frames
TBCs, VTRs, and production
COMPUTER GRAPHICS FOR
VIDEO   120
Originating computer graphics
Interface between people and
machines
CHARACTER GENERATORS   122
CREATING IMAGERY AND
EFFECTS   124
Computer generated imagery
(CGI)
Digital video effects
PROBLEMS OF CAMERA PICKUP
TUBES   126
Electronic problems
Visual problems
CHARGE COUPLED DEVICES   128
CCD layout and operation
Broadcast-quality requirements
CCD production problems
CCD ADVANTAGES   130
CCD field cameras
Economic advantages of CCDs
Visual advantages of CCDs
COMPOSITE VERSUS
COMPONENT VIDEO   132
Problems of composite video
Component video
Y/C
COMPLETE DIGITAL SYSTEMS   134
DIGITAL VIDEOTAPE
RECORDERS   136
D-1 component
D-2 composite
HIGH-DEFINITION TV   138
PATCH PANELS   140
What patch panels do
Patch panel components
Patching example

PATCHING RULES AND
PROCEDURES   142
  Patching rules
  Patching procedures
EXPLANATION OF A SMALL
PATCH PANEL   144
  Normal setup

SIMPLE PATCHING EXERCISES   146
    Summary of patch panels

FURTHER READING   149
GLOSSARY   151

Introduction

This is not a TV production textbook, but it is for TV production people. This book doesn't deal with TV production techniques or TV audio. There are already several fine books that cover those subjects. I see no reason to write what others have already done a better job of writing. Rather, this book deals with two interrelated subjects. I hope to show you how the various pieces of video equipment are integrated to form a complex video system. But to understand that, you must first have some knowledge of how the equipment works and what goes on inside it.

As TV equipment becomes more complex and sophisticated, it becomes more important to understand how that equipment works. This is especially true in the worlds of instructional and industrial TV, where one person may have to do it all.

Having an understanding of how the equipment and systems work gives you two distinct advantages. First, you will be more adaptable with different makes, models, and features. New buttons and knobs aren't as likely to intimidate you. Second, you can be more creative. You're not limited to what you have been shown, but can figure out new applications and how to solve problems for yourself.

There's a third advantage that some may debate. If you can understand and speak some of the engineer's language, you're more likely to turn those irascible souls into allies.

One need not be an engineer or know advanced math and physics to understand the basics of how TV equipment works. After teaching much of this material for more than ten years, I know that students with little or no math and science background can be taught to understand the equipment. However, students must realize that, because of their lack of background, they may not be able to get detailed answers to all of their questions.

Based on feedback from former students and their employers, I'm confident that the material in this book is the single most important body of material that my TV students receive. Many of them feel that it's even more important than experience using the equipment.

Since this book deals with television systems, it's difficult to understand some topics without the proper foundation. This book, then, uses the building block approach. Many topics rely on information from previous topics. Much of this book should be studied sequentially.

I've attempted to make this book as easy to read and understand as possible, but it should be recognized that few of us can learn TV production or TV systems just from a book. The best learning will take place if you use this book in conjunction with hands-on TV production experience. Everything from directing to shading and tape operation will help make the contents of this book more meaningful. I've tried to make the text clear, concise, and conversational in nature.

I have, on occasion, simplified the facts and various theories somewhat in order to make some concepts a little clearer. I hope more knowledgeable readers will not oppose these changes.

Acknowledgements

The contents of this book have been accumulated over a period of years. Much of this technical information was generously supplied by patient engineers who were willing to take the time to share their field with one who was less knowledgeable.

Among them were the engineering staff at the TV studio of California State University–Chico who, in my student days, were always willing to answer the technical questions I had that weren't covered in class. The man who designed and supervised the building of our studio at Cuesta College, Darrell Wenhardt, SAIC Broadcast Systems Integration of San Diego, always found the time to discuss and explain emerging technologies, even years after his contractual obligations to the school had ended. Finally, I've been fortunate with regard to the maintenance engineers I've known at Cuesta College. Ken James, now international products marketing manager in the Modular Products Division of the Grass Valley Group, and Jan Schaafsma, now a development engineer at Harris Corporation, spent many hours of discussion helping me to expand my knowledge of this field.

A number of people also read this manuscript and made suggestions, both major and minor, that have contributed to this project. Ravel Bastien, cofounder of Alpha Video Productions of San Luis Obispo, California, and Ken Zigler, owner of Aris Teleproductions of Arroyo Grande, California, offered the perspective of students who had been through my classes and who were now professionals in the field. Again, Darrell Wenhardt of SAIC Broadcast Systems Integration came through with many appreciated comments. I would especially like to thank Professor Donald R. Mott of Butler University for his careful reading of the manuscript and his detailed comments.

The Atom and Electricity

In order to understand the technical aspects of television systems and equipment it is important to know some basic theories, terms, and abbreviations in the field of electronics.

The parts of the atom
The first thing to do is get a basic understanding of how electricity works. If you go back to your high school science classes, you'll recall that an atom is made up of three parts. The neutron is in the center or nucleus of the atom and has no electrical charge. Protons, also in the nucleus of the atom, have a positive charge (+). Like the planets circling the sun in the solar system, *electrons* circle the nucleus and have a negative charge (−). Since there are the same number of protons and electrons, the atom as a whole has no electrical charge.

The flow of electrons through metals
In some elements, usually the elements that we call metals, electrons can be very easily dislodged from their orbits. When they are knocked out of their orbits, they are attracted to other atoms and knock electrons from their orbits. This flow of electrons is electricity. Of course, the atoms that have lost electrons now have an overall positive charge and tend to attract the loose electrons.

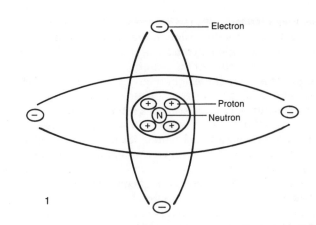

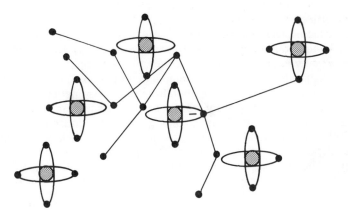

1. The parts of the atom.

2. The flow of electrons.

All complete circuits are loops of flowing electrons.

Basic Circuits

Direct current (DC)

A flashlight battery has the ability to provide a flow of electrons, but accomplishes nothing sitting on the shelf, since it is only stored energy. Only when the positive terminal of the battery is attached to one end of a light bulb and the other end of the light bulb is attached to the negative terminal of the battery do we have a completed circuit, and the light bulb emits light. In this circuit all of the electrons continue to flow in the same direction. This is called *direct current* (DC). Of course, if there's only one wire between the light bulb and the battery, or if there's an open switch, the circuit is incomplete. The electrons have to flow through a complete loop in order to do work.

Such is the case with all electrical circuits. There must be a complete loop providing both a place for electrons to come from as well as a place for them to go. This is why the plugs to all your appliances at home have two prongs. Most electrical devices will also have a switch somewhere in the circuit to allow you to interrupt the flow of electrons.

Alternating current (AC)

The preceding example explained a simple DC circuit. Most circuits used in video equipment use direct current. However, there is another type of current that you will commonly encounter called *alternating current* (AC). With AC the flow of electrons changes direction constantly. The current flows from negative to positive, then from positive to negative, and so on. AC is much more efficient for transmission through wires over long distances. That's one of the reasons that it is used for household electricity. Some household appliances work better on AC (your clothes washer for example) and others work better on DC (the internal circuits of your TV), so the ability to easily change AC into DC is another good reason why we use AC power for our main electricity supply.

1. Battery (stored energy).

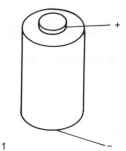

2. Simple circuit.

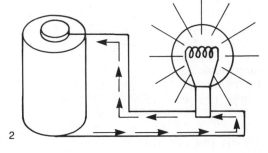

3. Simple circuit with switch.

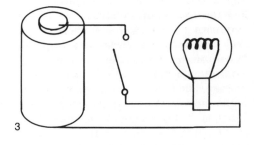

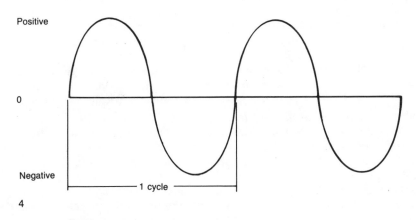

Positive

0

Negative

1 cycle

4

4. Alternating current.

When you're dealing with electricity, you need to be able to measure it.

Units of Measurement (1)

Voltage

There are several areas of measurement relating to electricity. The first of these is *voltage,* which is measured in *volts* (V). Voltage is the pressure of the electricity. In a given medium, the speed of electricity is constant, but the pressure is not. The typical flashlight battery has a voltage of 1.5 V. The electricity in your home, on the other hand, is 120 V. So you could say that the electricity in your house has a lot more pressure behind it.

Current

The second area of measurement for electricity is *current,* which is mea-sured in *amperes* (amps). Current is the volume of electrons, that is, the number of electrons passing a certain point in a given time. A current of 4 amps has twice as many electrons passing by as does a current of 2 amps.

Power

Voltage and current together determine *power.* Power is a measurement of work being accomplished and is measured in *watts* (W). Watts are de-termined by multiplying V by amps. For example, if you have a light bulb that draws 1.25 amps and the house voltage is 120 V, your light bulb is a 150-W bulb (1.25 amps $\times$ 120 V $=$ 150 W).

Resistance and impedance

Another area of measurement is *resistance.* Any DC electrical circuit will resist the flow of electrons. We measure this resistance in *OHMS* (Ω). Closely related to resistance is *impedance.* Impedance can help tell the production person if two or more circuits will interact well. The following oversimplified example may help you understand the concept. If your stereo amplifier has a speaker impedance of 8 Ω, this means that it is designed to hook up to speakers that have 8 Ω of resistance. If you connect your 8-Ω amplifier to your 8-Ω speakers, everything works great. But, what happens when you connect that 8-Ω amplifier to speakers that have 10,000 Ω of resistance? Not much! The system just isn't designed to overcome that much resistance. On the other hand, if you have both an amplifier and speakers with 10,000 Ω of impedance, everything works just fine. But if you connect that 10,000-Ω amplifier to speakers that have 8 Ω resistance, you've got problems. You could destroy your speakers! They're just not designed to work with that amplifier. You have what's called a *mismatch.* Impedance is an important factor when integrating electrical components.

Mathematical symbols and formulas

Because they all concern the flow of electricity through a conductor, these basic units of measurement are all mathematically related. In addition, when working with units of measurement mathematically, we give them different symbols. The mathematical symbols and the basic formulas are shown in the table. You may not need to memorize these formulas, but you should know that they exist and that the units are interrelated. It is also important to remember the different mathematical symbols.

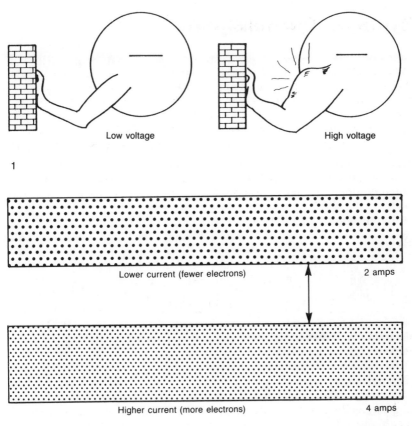

Low voltage

High voltage

1

Lower current (fewer electrons) 2 amps

Higher current (more electrons) 4 amps

2

1. **Voltage**.

2. **Current**.

Unit	Mathematical symbol	Basic formulas
Watts	W	$E \times I$ or $I^2 \times R$ or $\dfrac{E^2}{R}$
Volts	E	$I \times R$ or $W \times R$ or $\dfrac{W}{I}$
Amperes	I	$\dfrac{E}{R}$ or $\dfrac{W}{R}$ or $\dfrac{W}{I}$
Ohms	R	$\dfrac{E}{I}$ or $\dfrac{E^2}{W}$ or $\dfrac{W}{I^2}$

There are also other measurements you need to know.

Units of Measurement (2)

Voltage, current, and resistance are basic measurements of electricity, but when you need to apply electricity to television, there are some other measurements you'll need to know. Among these are frequency, hertz, and AC frequency.

Frequency
Frequency is an action that repeats itself. If you have an electrical circuit that puts out repeated and equal bursts or pulses of energy at 100 of those pulses a second, the frequency of that circuit is 100 pulses per second. But we measure frequency in hertz (Hz) so the frequency is 100 Hz.

AC frequency
In our earlier discussion of AC you learned that the flow of electrons in AC current constantly changes direction. If the electricity in your home is 120 V AC, what happens is that the electricity goes from 0 V up to +120 V, back down to 0 V, continues down to −120 V, and then goes back up to 0 V. This alternation between 120 V of positive electricity and 120 V of negative electricity is one cycle. Your household electricity does this 60 times a second. So the frequency of your household electricity is 60 Hz. Thus to be fully descriptive of the electricity in your house, you would say it's 120 V 60 Hz AC.

Capacitance
Another term you might hear the engineering staff of the station discussing is *capacitance.* A capacitor is a device that can store an electrical charge and the storage of that charge is called capacitance. The unit of measurement for capacitors is called the *farad* (F). Capacitors are used for the filtering of signals and they are used in power supply circuits to help in the conversion of AC to DC.

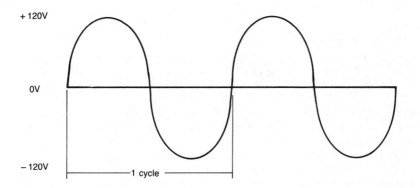

Alternating current.

Abbreviations

Kilo. *K* is the abbreviation for kilo, which equals 1,000. If something has a frequency of 10 KHz, it has a frequency of 10,000 Hz. You can just replace the word thousand with kilo. Likewise, if you have a light in your TV studio that uses 1 KW of power, it is using 1,000 W of power.

Mega. *M* stands for mega, or 1,000,000. So a generator that puts out 1 MW of power puts out 1,000,000 W of power. If your favorite radio station has an assigned frequency of 96 MHz, it has a frequency of 96,000,000 Hz.

Giga. *G* stands for giga and is equal to 1,000,000,000. So a measurement of 6 GHz would be equal to 6,000,000,000 Hz.

K, M, and G are used for units that are greater than one. There are, however, several abbreviations that are used for measurements that are smaller than one.

milli. The first of these is **m**, which stands for milli and means 1/1,000. If you took a measurement and it read 5 mV, that is read as five millivolts or five one thousandths of a volt. A measurement of 321 mA would be read as 321 milliamps or 321 one thousandths of an amp.

micro. The next abbreviation is μ, which stands for micro and means 1/1,000,000. Thus, if you have a reading of 25 μsec, that is read as 25 microseconds or 25 one millionths of a second.

nano. The last abbreviation is *n,* which stands for nano and means 1/1,000,000,000. So a measurement of 63 nsec would be read as 63 one billionths of a second.

Conversions

You also need to be able to convert these abbreviations. For example, you need to be able to convert KW to MW and vice versa. It's really a simple task, but there are a couple of rules that will make it easier.

To convert 2,575 KW into MW, first ask, "Am I going from smaller units to larger units, or from larger units to smaller units?" Since KW are smaller than MW, you're going from smaller to larger, therefore you move the decimal point to the left. How far to the left? Looking at the top chart at right you will find the difference in the number of zeros between K and M, and that's how many spaces you move the decimal point to the left. Since M has six zeros and K has three zeros, we move the decimal point three places to the left. As a result, 2,575 KW becomes 2.575 MW. Going from larger to smaller units works just the opposite way. This process is summarized by the charts at right.

```
G = giga   =   1,000,000,000
M = mega   =       1,000,000
K = kilo   =           1,000
    units  =               1
m = milli  =           1/1,000
μ = micro  =       1/1,000,000
n = nano   =   1/1,000,000,000
```

1

Smaller to larger, decimal point goes left.

Larger to smaller, decimal point goes right.

The difference in the number of zeros determines the number of spaces the decimal point is moved.

2

1. **Chart of abbreviations.**

2. **Rules of conversion.**

Induction and Noise

Induction

There are two other basic theories necessary for understanding television equipment. The first of these is *induction.* Any electrical circuit that has a changing flow of electrons will create an electromagnetic field around itself. For example, if you turned a flashlight on and off several times, the flow of electrons would be starting and stopping and a small electromagnet field would be created. However, if you left the flashlight on, the flow of electrons would be continuous and unchanging and there would not be an electromagnetic field. Since the flashlight in the first example uses very small amounts of electricity, its field would be very small — almost unmeasurable. But a high-tension power line running cross-country has an extremely strong electromagnetic field. When another circuit is placed within this electromagnetic field, a signal from the more powerful circuit is forced into the weaker circuit. The signal may take the form of static, as when you try to play the AM radio in your car near high power lines, or it may be actual information, as when you sometimes hear very weak background voices on the telephone.

Noise

Another thing that can create problems is *noise.* To see what noise looks like in video, unhook the antenna and/or cable from your TV. Turn your TV on. What you see is noise! If you happen to be near a transmitter and have your TV tuned to its channel, you'll also see some picture. This noise is obviously an undesirable feature. Too much of it and it interferes with the picture or signal. Inherent in every electrical circuit is a certain amount of this noise. If there is too much noise, then there is a problem. Certainly, if you want to watch TV, you don't want to see any apparent noise.

Signal-to-noise ratio

You need to be able to measure the relationship between the strength of the signal and the amount of noise the circuitry creates. This measurement is called the *signal-to-noise ratio.* We use the decibel (dB) scale to measure this relationship. The dB scale is a logarithmic ratio. The signal-to-noise ratio is doubled for every 3-dB difference between the strength of the signal and the strength of the noise. For example, if the noise in our system is 0 dB and the signal is 3 dB, then the signal is twice as strong as the noise; if the signal is 6 dB, then it's four times as strong as the noise; if the signal is 9 dB, it's eight times as strong; 12 dB, 16 times as strong; 15 dB, 32 times as strong and so on. In video, we like to have a signal-to-noise ratio of at least 45 dB.

Sometimes a signal-to-noise ratio is written as −45 dB. It really means the same thing. A 45 dB ratio means that the signal is 45 dB stronger than the noise; a −45-dB ratio means that the noise if 45 dB weaker than the signal — same thing. In real numbers a signal-to-noise ratio of 45 dB means that the signal is over 33,000 times stronger than the noise! That's pretty impressive.

1. Induction.

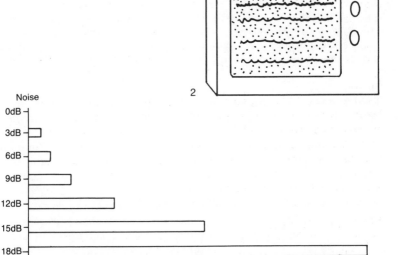

Electromagnetic field created by transformer

Portable radio in electromagnetic field

Static (noise) coming out of radio caused by transformer's electromagnetic field

1

2. Video noise.

2

Noise

0dB –
3dB –
6dB –
9dB –
12dB –
15dB –
18dB –

Signal

3

+ 45-dB signal to noise ratio = signal is 45dB stronger than the noise

– 45-dB signal to noise ratio = noise is 45dB weaker than the signal

Therefore, + 45-dB signal to noise ratio = – 45-dB signal to noise ratio

4

3. Signal-to-noise ratio. Comparison of the strength of the signal with the noise using the dB scale.

4. Minus and positive signal-to-noise ratios.

Camera System Components

There are two general types of cameras available: the studio camera and the field camera. Studio cameras are normally camera systems made up of two separate units: the *camera head* and the *camera control unit* (CCU). The camera head actually changes the light images into electrical signals. The CCU allows the engineer to make adjustments and control the quality of the picture while the camera is in operation. Field cameras, on the other hand, have everything combined into one unit. Any adjustments needed must be made before shooting begins.

Pickup tubes
Both types of cameras are built around a key component called the *pickup tube*. (Many of today's cameras use an imaging device called a CCD. This subject is discussed later.) There are many different types of pickup tubes, but they all operate in the same manner. The pickup tube is made of a glass enclosure that has the air taken out of it to create a vacuum (commonly known as a vacuum tube). At the front of the tube is the *target*.

Target
The target is made of a special material that has a specific reaction to light. When light strikes the target, an electrical charge is created. The brighter the light, the greater the charge. Thus, if you focus a picture of a person in a white shirt and black pants on the target of the tube, the white shirt will cause a greater electrical charge to form than the black pants will.

Cathode
In the back of this pickup tube there is a *cathode*. When this cathode is heated, it emits a stream of electrons (a heated cathode is also called an electron gun). Since the electrons have an electrical charge, they are controlled relatively easily with magnetic fields. This is why the camera pickup tubes will have massive coils of wire (electromagnets) around them. These are called *deflection coils* and allow us to steer the electron beam. The controlled beam hits the target and sweeps across it. As it does, it releases the built-up charges from the target. As these charges are taken off of the target they are amplified and fed into the camera's processing circuitry.

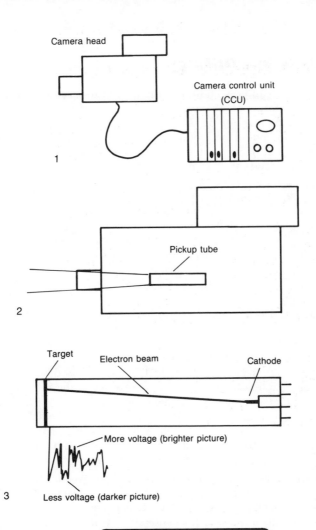

Camera head

Camera control unit
(CCU)

1

Pickup tube

2

Target   Electron beam   Cathode

More voltage (brighter picture)

Less voltage (darker picture)

3

Pickup tube   Deflection coils

4

1. **Studio camera system.**

2. **Pickup tube in a camera head.**

3. **Pickup tube.**

4. **Deflection coils around pickup tube.**

An electron beam sweeps its target to create video frames.

Camera Scanning

When a camera captures an image, the lens of that camera focuses the image onto the target of the pickup tube. Because of the way lenses work, the image will be inverted on the target. So if you are shooting a picture of a person standing on the ground, that person's feet and the ground will be focused on the top part of the pickup tube's target and his head will be at the bottom. However, the image itself is always scanned from top to bottom, left to right. In this case, the head of the person will be scanned first. Thus, even though in reality the electron beam sweeps the target from right to left, bottom to top, from the perspective of the image, the beam still sweeps from left to right, top to bottom. To simplify things, we will consider the scanning process from the perspective of the image.

If we aimed the camera at a picture of progressively brighter bars, from black to white, the black would cause the pickup tube to generate a small charge; the next bar, a brighter gray, would result in a stronger charge; the next bar an even stronger charge, and so on.

Interlace scanning.   Since a beam of electrons isn't very wide, one sweep across the target doesn't provide much information. To get more information, the electron beam has to make successive sweeps of the target. In the American system, it makes 525 sweeps or *lines,* to cover the entire target. But things aren't quite that simple. Rather than simply sweeping all 525 lines at once, the system sweeps the odd lines (lines 1,3,5,7,9, . . .) first and then sweeps the even lines (lines 2,4,6,8, . . .). Thus there are two separate fields of 262.5 lines each. These two fields are combined to form a single video frame of 525 lines, a complete picture. The electron beam scans so quickly that there will be 60 fields and 30 frames every second. This process is called *interlace scanning.*

Random interlace.   There are two types of interlace scanning. The first is called *random interlace.* In random interlace, the exact position of each line varies with each frame scanned. Certainly line 5 will be between lines 4 and 6, but it may not be exactly centered between the two lines and its position may vary a little with each frame. This is the system used with home video cameras.

Positive interlace.   The other type of interlace scanning is called *positive interlace.* In this method, each line has a specific position where it will be in every frame. So line 1 is in the exact same position in every frame, as are line 2, line 3, and so on. In positive interlace scanning each line will always be exactly centered between the lines that precede and follow it. All professional broadcast equipment uses positive interlace scanning.

You might wonder if you can see the difference between random and positive interlace. If you were to put the two systems next to each other on home TV systems, you probably wouldn't see any difference. The difference is important when you begin building a system and hooking various components together and trying to use sophisticated production techniques.

24

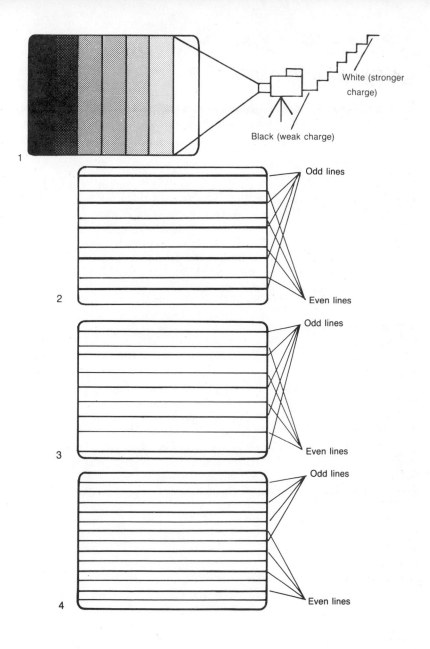

White (stronger charge)

Black (weak charge)

Odd lines

Even lines

Odd lines

Even lines

Odd lines

Even lines

1. **Brighter picture equals higher voltage.**

2. **Interlace scanning.**

3. **Random interlace.**

4. **Positive interlace.**

Blanking

The intensity of the electron beam during the scanning process is not constant. It varies in a logical, consistent pattern corresponding to the beam's movement across and up and down the pickup tube's target.

Horizontal blanking

In the scanning process, the electron beam scans the target, "reading" the first line of information. When the beam reaches the edge of the area defining the TV screen it is turned down to a low voltage, although it continues for a moment in the same direction. When the beam reaches the edge of the target, it quickly reverses direction and returns to the other side of the target. Once it reaches the other side, the beam resumes its original direction. Its voltage then is turned back up to scan another line. This process occurs every time lines are scanned.

The duration of the lowered voltage, from the end of one line to the beginning of the next, is called the *horizontal blanking.* During this horizontal blanking period, the return of the electron beam from one side of the target to the other is called the *retrace.*

Vertical blanking

You've seen what happens at the end of each line. Something similar happens at the end of each field. For a standard camera pickup tube, the electron beam is turned down to a low voltage before it retraces back to the top of the image. Once in position at the top of the image, the beam's voltage is turned back up and it starts scanning a new field. The time that the beam's voltage is turned down until it is turned back up again is called *vertical blanking* or the *vertical interval.* When the electron beam is retracing back to the beginning of a new field, it's called *vertical sync.*

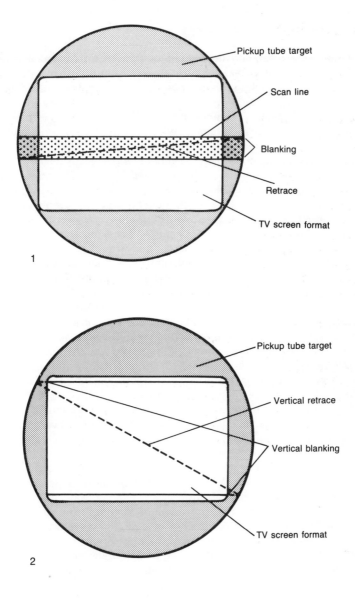

1. **Blanking and retrace.**

2. **Vertical blanking.**

Waveform Display

The horizontal scanning movement of the electron beam can be diagramed another way. Take a look at the figure at right. This graph represents an electron beam scanning one line of information, returning to the other side of the target during blanking, and reading a new line of video. Everything above the line represents the electron beam sweeping across the target in its normal direction, collecting information; anything below the base line is the beam returning in the other direction.

The graph actually measures the voltage of the electron beam in special units, called IRE units (for Institute of Radio Engineers, now called the Institute of Electrical and Electronics Engineers). The baseline is 0 units, which is simply 0 voltage. Notice that the height of the graph ranges from 100 to about −40 IRE. Just above the baseline are registered the black portions of the picture. This area near the baseline where the black portions of the picture appear is called the *pedestal*.

Above the base line, 100 IRE units indicates the maximum voltage the video system can handle and still provide a good picture. Information located here is called the *white peak*. Moving from left to right on this diagram, the voltages vary a great deal; as discussed earlier, the higher voltages are the bright parts of the picture and the lower voltages are the darker parts. Then there is a flat line of very little voltage; this is the start of blanking. The blanking continues for a little way then suddenly drops below the baseline; this is the start of retrace. After retrace, the voltage rises back above the baseline and remains at this low level until a new video line begins.

When you walk around the equipment area of a TV studio, you'll see several displays like the one just discussed. These are *waveform displays* and they're very important to both production people and engineers. The *waveform monitor,* which shows these displays, provides a graphic display of the black and white portion of the picture. The voltage signal generated from the black and white portion of the picture, its brightness, is also called the picture's *luminance.*

The waveform monitor's display will be similar to the simple two-line graph already discussed. Normally, however, the waveform monitor will continuously display the scanning of the electron beam across the target at the 30-frame-per-second rate. The blanking should remain the same from line to line, but the video information will vary as the beam scans higher and higher across the picture. Some monitors allow you to isolate one or two particular lines or points on the graph.

When shown on the waveform monitor, however, the blanking and retrace are called by different names. The retrace is called the *sync, sync pulse,* or *horizontal sync;* in fact, it's hardly ever called retrace around the studio. Two other parts of the blanking also pick up new names. The first part of blanking is called the *front porch,* and the last part of the blanking is called the *back porch.*

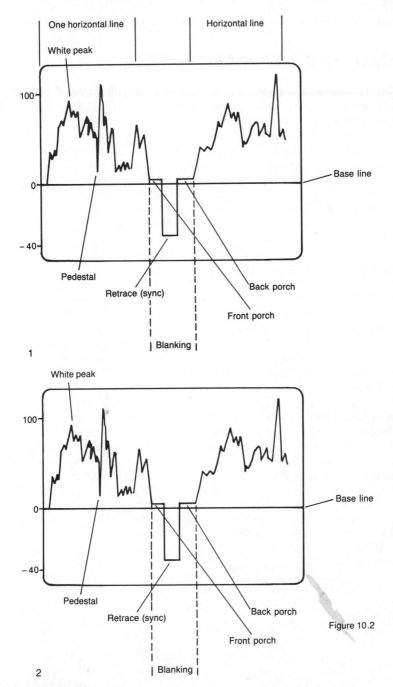

Figure 10.2

1. Waveform scan.

2. Waveform monitor continuously displaying the scanning of the target by the electron beam.

29

Cathode Ray Tubes (CRTs)

A television picture tube is properly called a *cathode ray tube* (CRT). The CRT is a large glass vacuum tube. The inside front of the tube is covered with a phosphorescent substance that glows when struck by a beam of electrons — the stronger the beam of electrons, the brighter the glow; the weaker the electron beam, the dimmer the glow. At the back of the CRT, in the narrow neck, is an *electron gun* (a cathode that is heated) that emits a beam of electrons — the higher the voltage activating the gun, the stronger the electron beam; the lower the voltage, the weaker the beam. The direction of the electron beam is controlled by the *deflection yoke* (a group of large electromagnets surrounding the middle section of the CRT).

CRT and pickup tube relationship

The processes of the CRT and pickup tube complement each other. The pickup tube in the camera converts the light images into the *analog video signal* (analog refers to a signal that communicates information through constantly varying electrical voltage), which is measured by the waveform monitor. The CRT converts the video signal to a pattern of lights, darks, and grays on its face that exactly duplicates the pattern on the target of the pickup tube. This relationship is illustrated by the figure at right.

Further, the motion of the electron beam within each tube is the exact opposite of the other electron beam. As previously explained, the electron beam in the pickup tube scans the target from bottom to top, right to left; the electron beam in the CRT sweeps across the tube from top to bottom, left to right. As previously discussed, this inverse action is necessary because the lens system of the camera focuses an upside-down, reversed image onto the target of the pickup tube, in the same way your image appears upside down when reflected in a shiny spoon. The right to left, bottom to top scanning motion of the pickup tube is inverted in the receiver so that the image appears correctly on the CRT screen. Again, both processes scan the image itself in the same way.

Need for interlace scanning

This is a convenient time to explain why the interlace scanning method is used in the system. Our TV system was developed over 40 years ago. Engineers had to work within the limitations of technology of the times. One of those limitations was the phosphor coating of the CRTs. When the CRTs electron beam struck a phosphor on the face plate, it caused that phosphor to glow; as soon as the electron beam left the phosphor, the glow started to get weaker. If our CRT scanned from line 1 all the way down to line 525 in succession, by the time the electron beam got to the bottom of the picture, the top of the picture would be pretty dark. To prevent this, the electron beam scans a field of 262.5 lines and then goes back to the top of the picture. Just as the lines at the top were starting to

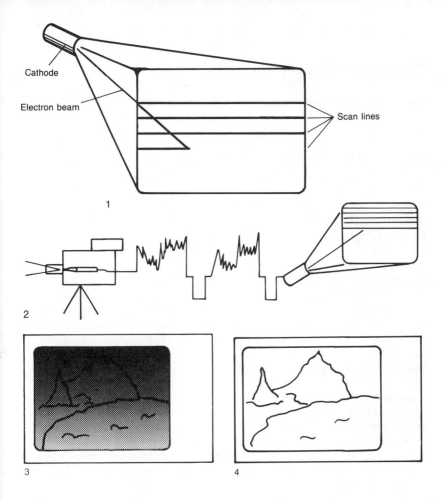

1. **CRT.**

2. **Analog video system.**

3. **TV without interlace scanning.**

4. **TV with interlace scanning.**

darken, the electron beam filled in the spaces between the darkening lines with bright new information. As a result, the overall picture maintained an even brightness. So, the interlace scanning system is the method used to ensure that the picture has an even brightness throughout instead of having separate bright and dark bands.

With today's technology it is possible to design systems that use sequential scanning. In fact, engineers would prefer to do this because it eliminates some problems created by interlace scanning. However, all current TV sets would be incompatible with such a system. As a result, we still use interlace scanning.

Color System

Color versus black and white

You've been shown how a visual image is scanned in a pickup tube, converted to an analog video signal, and transferred to a CRT for display. Black and white video systems operate in the way already discussed: the camera has one pickup tube, and only black and white information — luminance — is recorded and conveyed to the CRT. Color video is based on these basic principles, but is a little more complex.

Additive and subtractive colors

Instead of having just one pickup tube, high-quality color cameras have three pickup tubes, one for each of the primary colors: red, green, and blue. Those of you who have had art classes might say, "Now wait a minute, the primary colors are red, blue, and yellow." Well, you're right; if you're dealing with *subtractive colors.* Subtractive colors are what you deal with if you're mixing paints; subtractive colors reflect light off themselves. In video, we're dealing with *additive colors.* Unlike subtractive colors, which depend on substances that interact with white light, additive colors depend on the color of the light itself. With additive colors, the primaries are red, green, and blue.

Complementary colors

You can see by the figure at right that by mixing what appears to be equal parts of any two primary additive colors you will get one of the *complementary colors;* that is, equal parts of red and green will produce yellow, red and blue will produce magenta, and blue and green will produce cyan. When all three primary colors are mixed together we have white light. By changing proportions of the mix, brightness (intensity), and saturation of the colors, an infinite number of colors in the visible spectrum can be produced.

Red, green, and blue tube differences

There is really little difference between the three tubes; they come off the same production line. Each generally can process equally well, whatever the color light it is ultimately "assigned" in the camera. This is assuming that a tube does not turn out more sensitive to one particular color due to occasional variances in production; the tubes are tested for this.

Between the lens and the pickup tubes there's a beam splitter that divides the light into its three components and directs each part to the proper tube. Beam splitters may use prisms or a group of special mirrors to split up the light into its various parts (white light is made up of 59% green, 30% red, and 11% blue). Thus, coming out of the back of each of the tubes is an analog signal of that part of the picture.

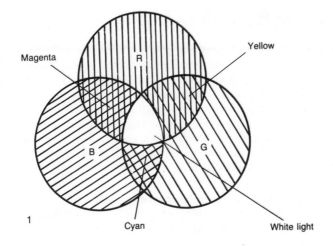

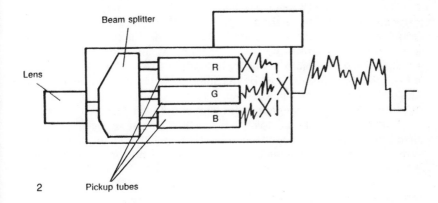

1. TV colors.

2. Color TV camera.

How the Eye Sees Light

In the section on complementary colors it was indicated that a mix of equal parts of red, blue, and green light would produce white light. Yet, it was also just asserted that white light is 59% green, 30% red, and 11% blue. How is this contradiction explained? It isn't really a contradiction. Your eyes are not equally sensitive to all colors of light. What looks like equal amounts of red, green, and blue isn't really equal amounts at all. The camera's pickup tubes, however, are equally sensitive to all colors of light. For your eyes to see white on the television screen, the TV camera must produce a picture that is 59% green, 30% red, and 11% blue.

Color temperature. If you've ever taken photographs outside and then gone indoors and taken more pictures without using a flash you've gotten some disappointing results. Chances are that the pictures taken outside looked great, but those taken indoors had an orange/yellow cast to them. What you've seen is the result of the difference of *color temperature.* Color temperature is a way of measuring the color characteristics of light. Color temperature has nothing to do with heat or cold. The higher the color temperature, the more blue there is in the light. The lower the color temperature the more orange in the light. The light outside on a bright, sunny day might be around 5,000° K (Kelvin, the scale used to measure color temperature). The light in your home is probably around 2,600° K. This is why the colors in the photographs look so different. If you use a flash indoors, the light will be the same color temperature as daylight and your photographs will look fine.

You don't see the change in color temperature because your eyes and brain compensate for you. If you see a person in a yellow jacket outdoors then the two of you go indoors, your brain knows that its the same jacket and it couldn't have changed colors. Your brain sees the jacket as being the same color. A camera can't do that.

Filters. What a camera can do is use filters. Most cameras have a built in filter wheel. This is a device that will allow you to place one of several filters between the lens of the camera and the beam splitter. Most cameras are set up to operate with TV studio lights, which have a color temperature of 3,200° K. If you go outdoors to shoot, you would change the filter wheel to compensate for the change of color temperature.

White balance. Changing the filter wheel is only the first step. Because clouds, shade, reflections, and other conditions all have an affect on color temperature, you will have to *white balance* your camera every time you set it up in new lighting conditions, which is simply establishing the proper color combination for given lighting conditions. To white balance a camera, simply aim and focus it on a white card and push the white balance button. The electronic circuitry of the camera will then adjust its

Overcast sky	7,000° K	Blue
Noon daylight	5,000° K	
TV studio lights	3,200° K	
Household lights	2,600° K	
Sunrise/sunset	2,500° K	
Candle	1,900° K	Orange

1

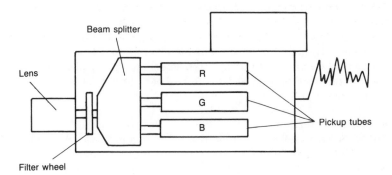

2

1. **Approximate color temperature of some common light sources.**

2. **Location of the filter wheel in a television camera.**

light reception so that the green tube produces 59% of the picture, the red tube produces 30% of the picture, and the blue tube produces 11% of the picture (see previous discussion). Since the camera is now properly mixing the three primary colors, the rest of the color spectrum will be fine.

35

Now you need to merge the signals.

Registration and Encoding

Registration

If you look closely at the drawing of the color TV camera, you might discern a potential problem. Once the light gets through the lens at the front of the camera, its components have to follow different paths, each of a different length. Since light and electricity move at constant speeds through a given medium, a longer path means it will take longer for the signal to go through that channel. Although we have only one picture going into the camera, we have the three components of that picture coming out of the back of the pickup tubes at different times. This is represented by the large Xs in the drawing, and would show up on the screen as up to three distinct offset images, each of a different color. When this happens, the camera is out of *registration.* Other registration problems are created by individual tube and scanning differences.

Fortunately, there are registration circuits that can be adjusted to bring the camera into registration. In short, adjusting the registration on a camera means ensuring that all three tubes put out the same picture at the same time.

Encoding

The three color signals are then combined in a process called *encoding.* The encoded signal (and the color signals, for that matter), now called a *composite signal,* actually is composed of two parts — the *chroma* (color) signal and the luminance (brightness) signal. Although there is some variation, the separate luminance signal is formed by skimming brightness information from each of the three tubes. It is then recombined with the chroma signal to create the composite signal. If you had only a black and white TV, you would receive and watch only the luminance signal.

The encoded, composite signal is fed out the back of the camera.

Home video cameras

It should be noted that most color home video cameras use only one pickup tube instead of three. These tubes have a filter of colored stripes in front of them. This striped filter breaks up the light into the three primary colors. Circuitry in the camera then combines these stripes of color into a composite color picture. As a result of this process, these cameras seldom approach broadcast quality. They rarely have the picture sharpness or color saturation needed for professional work. Therefore, these cameras are of little relevance to a discussion of broadcast television.

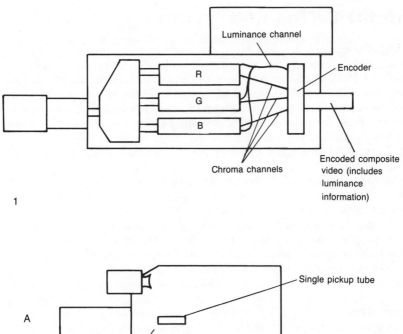

1

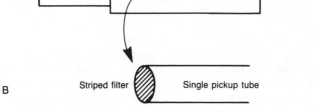

A

B

2

1. **Encoding.**

2. **Color home video camera.**

Color CRTs

Just as the color camera works in a manner similar to the black and white camera, the color CRT works in a manner similar to the black and white CRT. The color CRT has three sets of color phosphor dots laid down on the inside of the CRT. These are laid down in a specific pattern. The exact pattern will differ from manufacturer to manufacturer, but for the sake of explanation, a very common pattern, the triad, will be used. As you can see from the first figure at right, the triad is made up of one red phosphor dot, one green phosphor dot, and one blue phosphor dot arranged in a triangular pattern.

At the back of the CRT there are three separate electron guns, one each for the red, green, and blue information. Close to the front of the tube is a thin metal mask. This mask has tiny holes in it, arranged in such a way that only the red electron beam can strike a red phosphor, the green beam a green phosphor, and the blue beam a blue phosphor. These phosphors are so small and close together that they can't be seen as separate and distinct unless they are looked at under magnification. Thus, when they are struck by the electron beams, their colors blend together to produce the same color that the camera split up. The red part of a stop sign, for example, would cause only the corresponding red phosphors to be illuminated while the white letters of the sign would cause all of those corresponding phosphors to glow.

Convergence

Just as there can be a registration problem with the cameras, there can be a corresponding problem with the CRT called *convergence.* If there is a convergence problem we could see up to three distinct offset images of different colors. Convergence shouldn't be a problem with most home TVs, but it can be a problem with some large-screen projection units. You can usually make some control adjustments to lessen or minimize convergence.

It can be seen from this discussion that the color video system is considerably more complex than a black and white system. In fact, the color camera might be thought of as three synchronized cameras and the color CRT as three separate CRTs working in exact registration together.

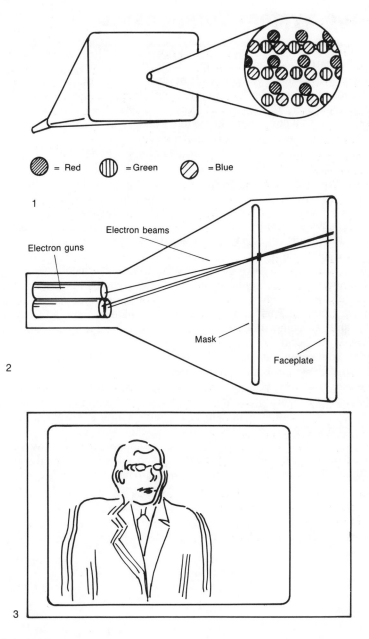

= Red   = Green   = Blue

1

Electron beams

Electron guns

Mask

Faceplate

2

3

1. Color CRT.

2. Components of a color CRT.

3. CRT convergence error.

Any practical video system needs at least four basic components.

Video System Components

Now that you have an idea of how cameras and monitors work, you're ready to start plugging those components into a system.

Any practical video system needs at least four components. The first is a camera to change the light images into electrical signals. The second component is a *sync generator,* which, as described later, establishes basic signals that coordinate the system's operation. In some systems, like home video cameras and professional field cameras, the sync generator is built into the camera head. The third component of a basic video system is a *videotape recorder* (VTR) on which to record the pictures. The possible exception to this requirement is if the video signal is directly transmitted via broadcast or cable, as in a live news report. In these cases a VTR is not essential — you don't necessarily have to record the signal — but in most cases it is still a good thing to have. The final component is a *monitor* to display the pictures the camera makes. These four components comprise a basic video system.

These components allow you to receive, coordinate, record, and review visual images. You can produce a videotape simply by hooking up these basic elements. However, the quality may not be very good, and rarely of broadcast quality. This is partly because of problems inherent in the imaging and recording processes, and partly because you have very little knowledge or control of the process.

Additional components, some of which I'll discuss here, can help you overcome these problems. Some devices, such as vectorscopes and waveform monitors, will help you identify potential equipment failures and keep the system operating correctly. Other components, time base correctors, for example, can help you minimize inherent problems in the production process and achieve a higher quality tape.

Finally, additional cameras, recorders, switchers (which allow you to channel the signals from several sources), and similar components will all contribute to a more complex system with greater potential.

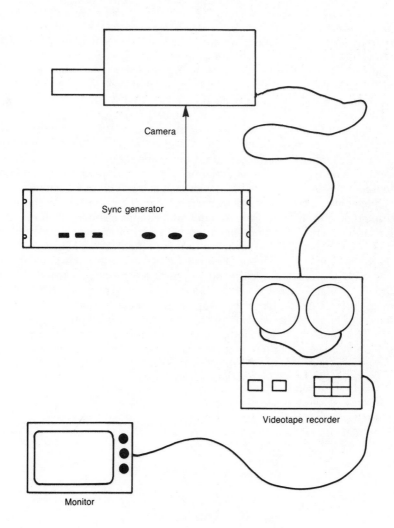

Camera

Sync generator

Videotape recorder

Monitor

Components of a video system.

41

The system needs to have a standard beat.

Sync Generators

What exactly is a sync generator? A descriptive name would be a *color synchronizing pulse generating system.* No wonder it's simply called a sync generator. The sync generator is the master clock that coordinates the whole system. Like a drummer in a band, it makes the rhythm that keeps everything in time and running together as a unit, rather than as a bunch of individual components each doing their own thing. When you're working with a single camera, it's not that big of a job, but when you start developing a more complex system, the job becomes more complex.

Imagine that you have two separate cameras working independently of each other. You turn on their power and off they go. The fact that the electron beams of their pickup tubes are scanning different parts of their targets is no big deal. But what happens if we try to hook those cameras together through a *switcher* (a device that allows you to instantaneously change between different video sources)? In the diagram at right, the electron guns of each camera are in different locations scanning their targets. Assume camera 1 is being shown on the *program monitor* (the monitor that shows what is being recorded or transmitted). What happens if you try to make an instantaneous change (cut) to camera 2? If the cameras are not synchronized, the picture on the program monitor will roll, jump, flicker, or tear — what's called a *glitch.* In other words, there will be a major picture disruption or breakup when the cut is made. Imagine, on the other hand, what would happen if the electron guns on the two cameras had been scanning together. Depending on the switcher, there might still be a breakup, but it would be less severe. The sync generator is the remedy to this problem. It keeps the cameras scanning together so that you can make clean transitions between cameras.

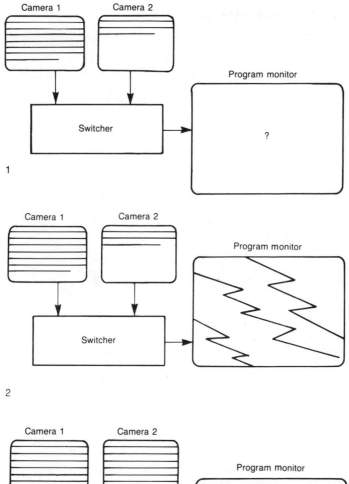

1. An unsynchronized system.

2. Making a transition with an unsynchronized system.

3. Making a transition with a synchronized system.

Sync Generator Signals

Drive pulses.   The sync generator provides four primary types of signals, and each of these signals has a couple of subtypes. The first thing the sync generator provides are *drive pulses.* These are the signals that tell the electron gun to scan across and down the target of the pickup tube. There are two types of drive pulses, one controlling each direction the electron gun scans: *horizontal drive pulses* and *vertical drive pulses.*

Blanking pulses.   The second type of signal the sync generator provides is *blanking pulses.* As you might suspect, the blanking pulses tell the electron gun when to lower its voltage at the end of a line or end of a field. There are two types of blanking pulses: *horizontal blanking pulses* (end of a line) and *vertical blanking pulses* (end of a field).

Sync pulses.   The third type of signal the sync generator provides is *sync pulses.* The first two types of sync pulses should be pretty easy for you to figure out. *Horizontal sync pulses* tell the electron gun to retrace to the beginning of the next line (these are the sync pulses seen below the baseline on the waveform monitor). *Vertical sync pulses* are at the end of a field and tell the electron gun to retrace to the top of the picture. The third type of sync that the sync generator generates is *color sync,* also called *color subcarrier, color burst,* or 3.58 (pronounced "three five eight," the meaning of which will be explained later).

Color burst.   Color burst is a reference signal that is inserted in the back porch of every line of video. This reference is used when encoding the color information with the luminance information. The color burst acts as a marker, and the color information for each line is encoded onto that line based on the marker. If the marker is off, so is the color. The key aspect of that marker is its frequency. The color burst is really a burst of energy at a specific frequency. That frequency is 3,579,545 Hz $\pm$ 5 Hz. This is normally rounded up to 3.58 MHz, hence the name 3.58. If that frequency is off, the colors will be off. This is, then, an extremely important signal the sync generator provides in a color system.

Combining sync with video

Take a moment and think about how the video system has been described here. The pickup tubes in the cameras are converting the light images into electrical signals, but the drive, blanking, and sync pulses that tell those pickup tubes to scan a line, turn down, retrace, turn back up, scan another line, start a new field, and so on are coming from outside the camera. Yet coming out of the camera, the video and sync information are combined. This is called *composite video.* Video information without the sync information is called *noncomposite video.* So when you look at the waveform monitor, you're looking at composite video. In a broadcast-quality system, video is always 1.0 V (140 IRE units) peak to peak (from the bottom of sync pulse to the top of the white peaks) across 75 Ω impedance. By looking at the bottom figure, you can see that the video portion is 0.7 V and the sync

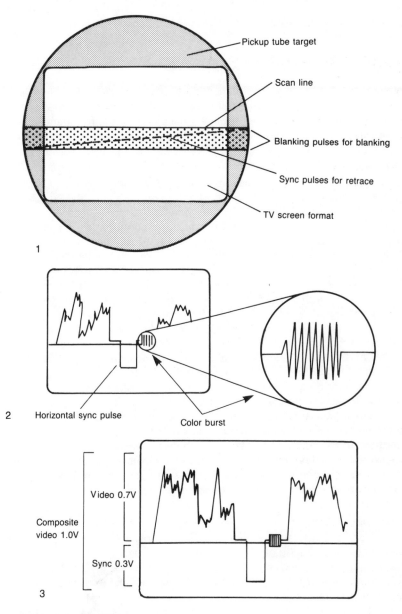

Pickup tube target

Scan line

Blanking pulses for blanking

Sync pulses for retrace

TV screen format

1

2 Horizontal sync pulse

Color burst

Composite video 1.0V

Video 0.7V

Sync 0.3V

3

1. **Parts of the TV picture.**

2. **Color burst in a waveform scan.**

3. **Components of composite video.**

portion is 0.3 V (these figures are rounded off; the video portion is 0.714 V and the sync is 0.286 V), adding up to 1.0 V. The entire video signal is less powerful than the energy from a flashlight battery.

The vectorscope enables you to monitor the color signals.

Vectorscope

While the waveform monitor presents a graphic display of the black and white information in a picture (its luminance), a vectorscope presents a graphic display of the color information in a picture (its chrominance), and allows us to adjust color levels.

Reading the vectorscope

Although the vectorscope is used primarily by engineers, production people must know the basics of reading and operating it. Like the waveform display, a vectorscope displays signal patterns for scan lines on a CRT. Both the waveform monitor and vectorscope continuously display 525 lines (one video frame) 30 times every second. Each can isolate specific lines. Since the phosphors on the faceplate of the vectorscope's CRT maintain an afterglow, we actually see the display of a great number of lines at once, producing a composite of the whole picture. The vector display is round, with both black and white located in the center. The little loop in the signal to the left of center (at the 9 o'clock position) is the color burst. The primary and complementary colors are assigned locations in relation to the color burst. Going in a clockwise direction, yellow is about 10° beyond the color burst, red 76° beyond the color burst, magenta 120° beyond the color burst, blue 190° beyond the color burst, cyan 256° beyond the color burst, and green 300° beyond the color burst. This assignment of information in relation to the color burst determines the color's hue. In the figure at the top you see the typical vector display. In the figure in the center you see what is really the same display, but it has been rotated 90° to the right. The relationship between the color burst (hard to see behind the heavier lines) and the individual colors is unchanged. If we tried to hook together cameras with these two displays in a system we would have a *phase* problem, a phenomenon to be explained later.

Color bar display

Color bars are reference signals produced by the sync generator and placed at the beginning of a tape when it is recorded. If we read the color bar display on the vectorscope, we can adjust the video system for proper color. Probably the only time production personnel will need to use a vectorscope is for adjusting color.

The sequence of colors in the color bar display is identical to the path of the signal displayed on the vectorscope; from center (both white and black) to yellow, cyan, green, magenta, red, blue, and back to center.

Look at the vectorscope diagram. You will notice that each color falls into its own little box on the vectorscope. This indicates that you are seeing the full level of *chroma.* Take the red bar, for example. If the chroma were cut in half, the trace on the vectorscope for the red color would stop about half-way between the center of the display and the red box. The basic color would still be red, but it would be a more pale red.

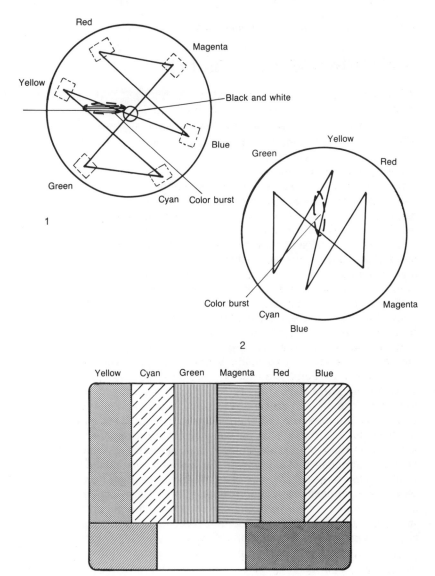

1. **Schematic of a vectorscope monitor.**

2. **Vector display rotated 90 degrees.**

3. **Color bars.**

When playing back a videotape, you'll look first at the display of the color bars and, if necessary, make a few adjustments to get the same pattern as in the first figure. Once the color bars are right, the colors in the video images that follow on the tape should accurately represent the colors in the scene that was originally photographed.

Sync Flow Diagrams (1)

Where do the various sync, drive, and blanking pulses go? The easiest way to determine this is by using flow diagrams. Flow diagrams are very handy things to be able to read. They use geometric shapes to represent various pieces of equipment, and lines to represent the wires that connect the equipment. Flow diagrams enable you to see how the various components of a video system are integrated electronically. Most studios are designed using flow diagrams since, if something needs to be changed, it's much easier to erase something on a diagram and draw a new line before a studio is built than it is to rip out a wall to add a new cable after the studio has been finished.

Take a look at the sync flow diagram. This diagram is an electrical map of a typical, large (three-camera) studio video system. It charts the electrical pulses going into the cameras, as opposed to the camera flow diagrams (discussed later) that trace the electrical flow leaving the cameras. In the lower right portion of the diagram is a box labeled "Sync Generator/ Video Test Set." This represents our sync generator. Across the top of the sync generator are the outputs: sync = horizontal and vertical sync, blank = horizontal and vertical blanking, s.c. = subcarrier (color burst, 3.58, etc.), H. Dr. = horizontal drive, and V. Dr. = vertical drive.

Distribution amplifiers

Follow the path of the vertical drive pulses. You'll see that the path goes straight up and then makes a sharp left turn and comes to a triangle that is labeled PDA 1. That stands for *pulse distribution amplifier.* A distribution amplifier is a piece of equipment that takes an input and gives you multiple outputs of that same input. So if we have a vertical drive pulse coming into the pulse DA, our diagram shows that we have six of the same vertical drive pulses coming out of it. This is very handy since, unlike home video and stereo, you can't use splitters and other such items because they degrade the signal too much. As its name implies, the distribution amplifier avoids this problem partly by amplifying the multiple outputs it produces. The first output goes to camera 1 to drive the electron beams of its pickup tubes. The second goes to camera 2 for the same purpose. Similarly, the third one goes to camera 3 for the same reason. The fourth ones goes to the film chain camera (to be discussed later).

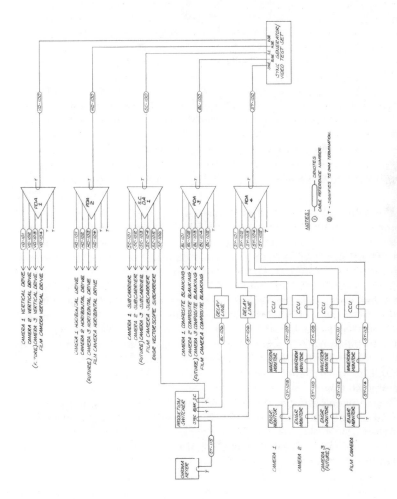

A typical sync flow diagram for a studio system.

49

Sync Flow Diagrams (2)

Termination

The fifth and sixth DA outputs have a T after them. This means they are *terminated.* But what does that mean? The circuitry of the DA is designed to match up with six pieces of equipment. But PDA 1 is only hooked up to four pieces of equipment; the other two outputs run up against a brick wall, so to speak. This could affect the four outputs that are hooked up. Some of the signal that was supposed to go to the last two outputs could bleed off to the first four outputs. As a result, the first four outputs could be changed from what they are supposed to be. To avoid this problem, a terminator is placed on each of the last two outputs. This makes the circuitry of the DA "think" there are pieces of equipment hooked up to it. Since all broadcast-quality video equipment has an impedance of 75 Ω, the terminator is little more than a 75-Ω resister hooked up to each of the final outputs. Many pieces of video equipment need to be terminated in this way. (In most newer DAs, however, there is sufficient isolation between outputs so that terminating unused outputs is not required.)

Subcarrier, blanking, and sync pulses

Going back to the sync generator and following the path of the horizontal drive pulses, you can see that it follows exactly the same type of path as the vertical drive. When you follow the path of the subcarrier, however, you see some differences. First off, our DA is an SCDA. That simply stands for *subcarrier distribution amplifier.* The first four outputs of the DA go to the same places as in the two previous examples, but the fifth output goes to the vectorscope. If you think about it, this makes sense. Since the vectorscope displays color information, might it not be able to make use of the color reference? Of course it can. You'll see that the last output of the SCDA goes to the production switcher. The production switcher can modify and do special effects with color, so it needs to have color burst to accomplish this.

The blanking output of the sync generator follows the same path as did the other outputs except that the fifth output goes to the production switcher. That's because the switcher has to have the blanking information in order to do its job properly. Ignore the delay line in that path for the moment; this will be discussed later.

The sync output of the sync generator follows the same path as the blanking did, but it's diagramed differently. In this case, the first output of the DA goes to the switcher, but this is not of great importance. Taking a closer look at the second output of the DA, you see that it goes to the camera 1 CCU, then from the CCU to a waveform monitor, and from the waveform monitor to an engineering monitor (labeled ENGR MONITOR). The waveform and engineering monitors are used by the video engineer (or shader) to control the quality of the picture coming out of the cameras. The third, fourth, and fifth outputs of the DA follow a similar path. What

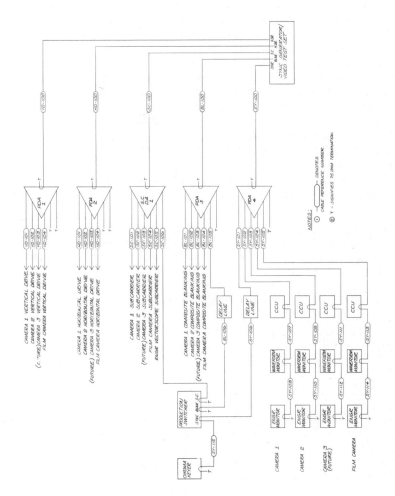

A typical sync flow diagram for a studio system.

these diagrams don't show is that the sync pulses also go from the CCU to the camera head, where they're used for the scanning processes. In addition, all the other sync generator outputs followed the same path: from the DA, to camera CCU, to camera head and waveform and engineering monitors.

51

Camera Flow Diagram

Take a look now at the flow diagrams for the cameras. As previously mentioned, this diagram traces output from the cameras to its ultimate destinations. The camera head is labeled "Studio Camera 1" and its output goes to its CCU. The CCU has outputs labeled "PGM" (program), "PIX" (picture), and "MON" (monitor). The PGM output is a fully encoded color output and goes from the CCU to a video distribution amplifier (VDA). The first DA output goes to the monitor bridge. These are the monitors that the director looks at to decide what picture to put on the air. The second DA output goes to the engineering (ENGR) switcher. This switcher allows the engineer to look at the various video sources to check for problems and fix them before they get too serious. The third DA output goes to the production switcher. This allows the technical director to record the picture that the director has selected. The fourth and fifth DA outputs are terminated, and the final DA output goes to the patch panel (the little circle with the number inside of it). Patch panels are like switchboards; they tie all of the system components together. Patch panels will be discussed later.

The PIX output allows you to look at the red, green, or blue channels independently or in any combination. You can see that this output goes to the waveform monitor, channel A. The waveform monitor has a switch that allows you to flip between two different inputs, and channel A is one of them. As you recall from the sync flow diagram, the output of the waveform monitor goes to an engineering monitor.

The MON output is the same signal as the PGM output. The MON output goes to channel B of the waveform monitor. So by flipping between channel A and B on the waveform monitor, you can compare the complete encoded signal (MON) with the outputs of the separate pickup tubes (PIX).

Camera 1 also has separate red, green, and blue outputs that go to the *chroma keyer.* The chroma keyer is special circuitry in the switcher that allows for some rather startling special effects. It will be discussed in more detail later.

Dropping down to camera 2, you see that it is just the same as camera 1 except that there are no outputs to the chroma key (camera 1 is our primary camera). Camera 3 is the same as 2.

Camera 4 is the film camera. It's used for transferring film to video and is much the same as the other cameras. In other words, this "film camera" does not record images on film. Rather, it helps transfer images already on film to video. Thus it has electrical inputs and outputs like the other video cameras. The first three outputs of the film camera DA are identical to those on the previous cameras. DA output 4 goes to the patch panel and output 5 goes to a monitor so that the film camera operator can see what the camera sees. The last output of the DA is terminated.

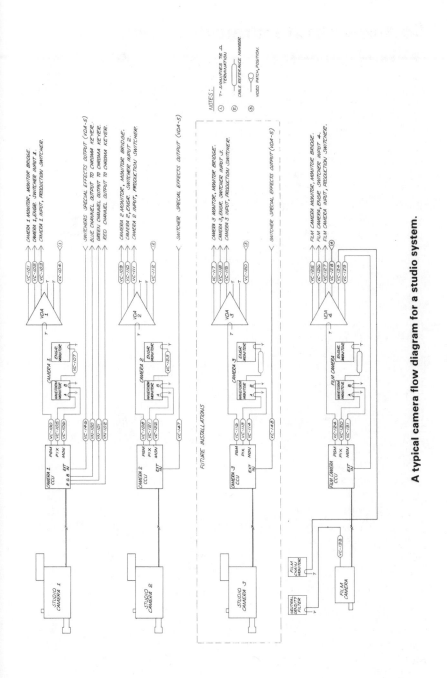

A typical camera flow diagram for a studio system.

53

Combining Sync and Camera Flow Diagrams

You can combine aspects of the two flow diagrams by eliminating a few parts and combining others to make it simpler. Follow the route of one of the sync signals; all of them follow similar routes. Following the color burst in the figure at right, you see that the sync generator goes to a DA. The first DA output goes directly to the production switcher. The second DA output goes to the camera 1 CCU. From the CCU it goes to the camera head and is combined with the video information. That composite video leaves the camera head, returns to the CCU, and from there goes to the switcher. The third DA output follows a similar path: DA to CCU, to camera head, back to CCU, and to the switcher.

Although only one specific system has been discussed, all systems are basically alike, whether using two cameras or 20. There will be differences, but the basic technique of running the sync through DAs to the various cameras and taking the composite video from the cameras through DAs to their various destinations will be similar in any professional situation.

Out of phase cameras

If you study the figure for a moment, you'll see that each of the paths is a different length. Since electricity travels at a constant speed, it will take the color bursts a different amount of time to follow each of the paths. Thus color bursts on each of the paths will arrive at the switcher at different times and the cameras will be *out of phase.* What happens if you're on camera 1 and you want to fade it down while you're fading up camera 2 (this is called a dissolve)? Which color burst will the switcher lock up to? It can only use one at a time.

The switcher will lock up on the color burst from camera 1, but as the dissolve progresses, the colors on camera 2 will look funny since they're referenced to the color burst of camera 1 instead of their own. When the dissolve is completed and camera 2 is fully "up," the switcher will lock to the color burst of camera 2 and the colors will snap back to normal. Fortunately, there are circuits in the cameras that can be adjusted to compensate for the difference in distances traveled by the color bursts. However, if the cameras are not set properly and are out of phase, the problem described above will result.

Timing the system

There can be problems with the sync and blanking signals similar to those possible with the color bursts. As is true of color bursts, sync and blanking signals from different sources can arrive at a common destination at different times. That's why there's a delay line between the DAs and the switcher to compensate for the problem. Adjusting the system to ensure that the blanking and sync signals from the various sources enter the

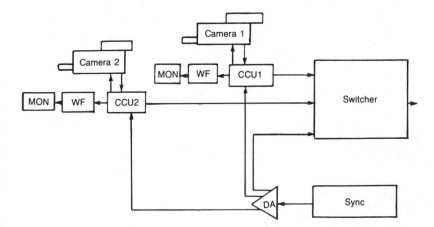

Combining sync and camera flow.

switcher at the same time is called *timing the system.* It's very important that the system be carefully timed when it's installed.

Many newer systems are using a different approach. Rather than distributing the various separate blanking and sync signals through individual DAs, a single synchronizing signal is sent. This signal is called *black burst.* Black burst originates from the sync generator. It consists of all the normal blanking and sync information along with black video. The individual pieces of video equipment will use the information from the black burst as a sync source. The process of locking up this information is called *gen lock.* This process greatly simplifies the system and reduces expenses.

Video dominates television, but film is still around.

Film-to-Video Transfer

Although today film is rarely used as a production medium at local stations, it will continue to be used for major series production and theatrical releases in the foreseeable future. Commercials, movies, and some syndicated shows continue to be distributed on film although more and more of them are coming out on tape. *Telecines,* or *film chains* as they're often called, are devices that convert film into a video signal by using a standard video camera. These are very basic and inexpensive film-to-video transfer devices, and will be used at schools and universities for years to come and will continue to be found at many local stations. There is also a great variety of more complex and expensive transfer devices. Because there is so much material currently being made and already out on film, such high-quality film-to-video transfer devices are in high demand in the major production centers.

Film is still the medium of choice for major big-budget top-quality production work. Most major national TV commercials, most major network TV series, and all major motion pictures are still done on film. The reasons for this are simple. Film gives a much higher quality picture. There's more detail on film, and film has a higher contrast ratio (it will handle a wider range between the bright and dark portions of the picture). Film also can give a wider variety of aesthetic "looks." Thus it is worthwhile for a station to have the ability to transfer film to video, and most stations do this by using a telecine or film chain.

A film chain can transfer both motion picture film and still transparencies (slides). You may find any of three motion picture formats available in a local station. *35 mm* film is the production medium of choice for TV series, national commercials, and major motion pictures. But since the telecine equipment in that format is very expensive, 35 mm is usually found only in large stations in the largest markets. *16 mm* film, on the other hand, is much less expensive (though still not cheap) and still produces good-quality pictures. Many motion pictures and national commercials are distributed to local stations on 16 mm film. This is the format that most TV stations have their film chains set up for. *Super 8 mm* is an inexpensive format, but the quality isn't very good. Despite the poor quality, because of the low cost you might find Super 8 film chains in very small stations and educational environments.

FILM FORMATS

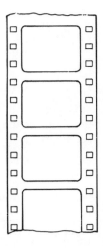

Super 8mm. Unprofessional in quality, but inexpensive

16mm. Minimum professional quality and moderately priced

35mm. Excellent quality, but very expensive

Film formats.

Film chains (telecines) transfer film to video.

Film Chain Layout

A telecine is really just a device that allows a special projector to project a film into the lens of a TV camera. At most local stations, however, it is set up in such a way that up to three projectors can be used with one TV camera. The first figure illustrates a standard film chain setup, assuming an overhead perspective so that you are looking down on the entire setup. As can be seen in the diagram, this setup uses two motion picture projectors and one slide projector. The keys to making this system work are the *multiplexer* and the special motion picture projectors. The multiplexer is in the middle of all the projectors and is little more than a group of mirrors that can be switched to direct any source within the film chain unit. In Figure 2 the mirrors are open so that the slide projector is projecting directly into the lens of the TV camera, which scans the image and converts it to video.

Condenser lens
You'll notice a couple of items located between the multiplexer and the lens of the TV camera. The *condenser lens* concentrates the image so that it will all enter the video camera lens. Figures 3 and 4 illustrate the action of the condenser lens.

Neutral density filter
The next item in the light path is the *neutral density filter* wheel. A neutral density filter reduces the amount of light passing through it without changing the color of that light. If you'll look at the camera flow diagram, you'll see that the neutral density filter wheel is hooked up to the back of the camera. If the image entering the camera is too bright, the filter wheel adjusts to reduce the light. If the image is too dark, the wheel adjusts to let more light in.

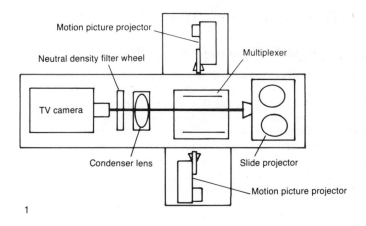

1

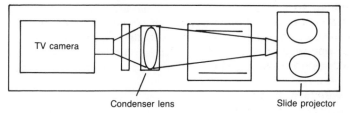

With condenser lens

2

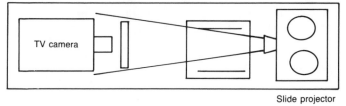

Without condenser lens

3

1. **Film chain diagram. The multiplexer is set for projecting slides.**

2. **Film chain with condenser lens.**

3. **Film chain without condenser lens.**

Film chain units require special, precisely aligned projectors and cameras.

Film Chain Projectors and Video Cameras

The figure at top right is another overhead view of a standard film chain setup. However, the position of one of the mirrors has been changed so that it reflects the projected image of a motion picture into the TV camera lens. You can't use just any projector for this purpose. You need a special film chain projector that is designed to synchronize with the TV system. Maybe you've seen movies of space launch control rooms with bank after bank of monitors. In those shots, there's almost always horizontal bars rolling through the monitor screens. That's because the monitors are out of sync with the camera taking the picture. The same effect would be seen without a special film chain projector. Motion pictures are shot at 24 frames a second and that doesn't mesh with TV's 30 frames. So, the film chain projector projects the first frame twice and the next frame three times, the next frame twice and the next one three times and so on. If you have 24 frames per second and half of them are projected twice and the other half are projected three times, that comes out to 60 projections a second and that matches nicely with 60 TV fields a second. With this system, then, there are no horizontal sync bars rolling through the TV picture.

The TV camera that receives these film images is essentially a standard TV camera. The projectors, multiplexer, condenser lens, and video camera are all very carefully aligned and locked into place. These alignments have very small tolerances. The slightest movement will throw an alignment off. For that reason, you must treat the setup very carefully, and never use a film chain as something to lean against.

The quality of the picture produced on the typical telecine unit is not the best. By the time the image goes through the projector lens, bounces off a mirror, passes through a condenser lens, a neutral density filter, and the camera lens, a lot of quality has been lost. However, many newer telecines are using charge coupled devices (CCDs; to be discussed later) rather than the traditional camera pickup tube. These CCD telecines provide a much sharper and a higher-quality image than was possible before.

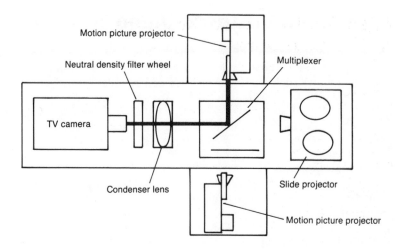

Motion picture projector

Neutral density filter wheel

Multiplexer

TV camera

Condenser lens

Slide projector

Motion picture projector

1

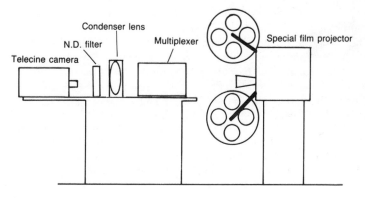

Condenser lens

N.D. filter

Telecine camera

Multiplexer

Special film projector

2

1. **Multiplexer set for projecting a motion picture.**

2. **Side view of a film chain setup.**

High-Quality Film-to-Video Transfer

There are, however, some transfer devices available that do produce an outstanding image. These devices are generally found in the major production centers, such as New York and Los Angeles. They will almost never be found in local stations. There are a couple of reasons for this. First, they are very expensive; second, they are self-contained units that will handle only one reel of film at a time (only one projector). However, they will transfer an exceptional picture to video tape.

Flying spot scanner

The most common of these high-quality devices use a *flying spot scanner.* That simply means that, rather than using a projector, they use a CRT to illuminate the film. The figure at right shows the basic setup of a flying spot scanner. Notice that the only lens in the system is between the CRT and the film. The electron beam scanning the blank CRT provides the illumination that is first picked up by the beam splitter and is then divided into its component colors. This is a much more efficient way to transfer film to tape and gives excellent results.

There can still be a problem with the frame rates, but the flying spot scanner can be set to scan each frame in the same multiples as the film chain projector did. In addition, some film directors who know they're going to use the unit will shoot their film at 30 frames per second in order to eliminate the conversion problem.

These new film transfer machines do a wonderful job, but their high cost, the fact that there's only one projector where a film chain has at least two projectors, and the fact that film is rarely used for production at local stations anymore generally will limit their use at local TV stations. Those stations will have to rely on the typical film chain until there are significant changes in the industry's choice of media.

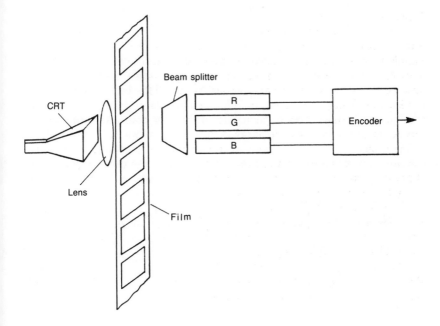

CRT

Lens

Beam splitter

Film

R

G

B

Encoder

Flying spot scanner.

Video Switchers

The *video switcher* is the keystone around which the rest of the TV studio is built. The switcher is a piece of equipment that allows you to choose from many incoming video sources and make transitions or other special effects between those sources. The sophistication of a switcher determines what transitions can be used between shots, what kind of special effects can be used, and how frequently they can be used. The capabilities of various switchers run from the very simple to the mind boggling, and their prices follow suit. Production switchers can cost from a few thousand dollars to several hundred thousand dollars!

Passive switchers

There are many types of switchers available. The simplest and cheapest of these is the *passive switcher.* The passive switcher is little more than an arrangement of mechanical switches. You push the button corresponding to a particular source and the picture changes. But when the picture changes you're very likely to get a glitch. Hence the common name "glitch box switchers." Passive switchers are totally unacceptable for production work. They might be found in places where clean transitions are unimportant, such as when an engineer needs to check the outputs of various sources.

Vertical interval switchers

The second type of switcher is the *vertical interval switcher.* This switcher has special circuitry that delays any cuts until the entire system is in vertical blanking — the vertical interval. This is one of the reasons why timing the system is so important (see page 54). Since vertical blanking happens 60 times a second, the delay is very small and imperceptible to humans, but it ensures sharp, clean cuts every time. Anything that is going to be switched for use on the air must go through a vertical interval switcher.

Component switchers

The next and newest type of switcher is the *component switcher.* Rather than using the complete encoded color signal, this switcher deals with the individual red, green, and blue components separately. It's almost like having three separate switchers combined into one package. The three color components travel through the switcher in parallel. This generally produces a much sharper picture and crisper special effects. These switchers do, however, cost a good deal more money than their encoded signal brethren do.

Special effects

Virtually all production switchers today come with special effects capabilities. How many of those special effects there are, what they are, how well they work, and how they can be used in sequences all affect the price of the switcher. The most common switcher effects capabilities will be discussed later.

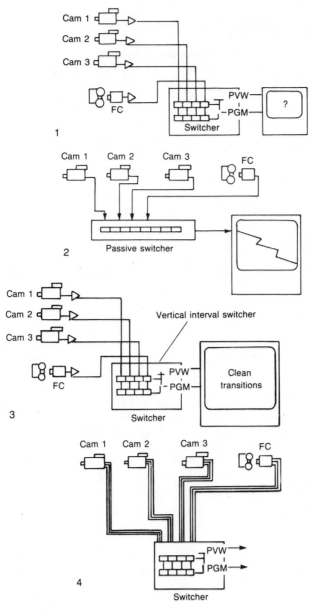

Cam 1
Cam 2
Cam 3
FC
PVW
PGM
?
Switcher

1

Cam 1   Cam 2   Cam 3   FC

Passive switcher

2

Cam 1
Cam 2
Cam 3
FC

Vertical interval switcher

PVW
PGM
Clean
transitions
Switcher

3

Cam 1   Cam 2   Cam 3   FC

PVW
PGM
Switcher

4

1. **Video system using a switcher.**

2. **Passive switcher.**

3. **Vertical interval switcher.**

4. **Component switcher.**

Switcher Applications

Production and editing switchers

There are three common situations in which a switcher will be used. The first of these is in production. Production here refers to creating finished video media that will ultimately be seen by a viewing audience: news, commercials, dramas, comedies, instruction, and just about anything else you can think of. These productions may require anything from the simplest to the most complex production switchers to accomplish the desired results.

On-air switchers

The *on-air switcher* generally coordinates sources of finished production and sends output directly to the transmitter. It will be switching between various videotape machines, film chains, network feeds, satellite feeds, and the studio. These switchers are almost always *audio-follows-video switchers.* That means that when the technical director pushes a button on the switcher, it changes both the picture and the sound. That's not the case with production switchers, where any changes in sound must be done separately. Since anything that normally goes to the transmitter has the picture and sound together, this makes things much quicker and easier for the technical director. On-air switchers usually have limited special effects capabilities. Because anything going to the transmitter is usually a finished product, there's little need for special effects at this stage.

Routing switchers

The final switcher application is for routing. Say that you're working in a large school system and you need to send seven or eight programs to different classrooms at the same time. You would use a *routing switcher* to accomplish this task. Routing switchers are often audio-follows-video units and they're often very large. The older ones are usually passive switchers. Some of the newer models use advanced electronics that allow for far greater flexibility using much less space.

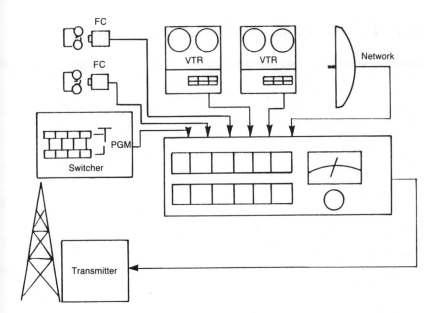

1

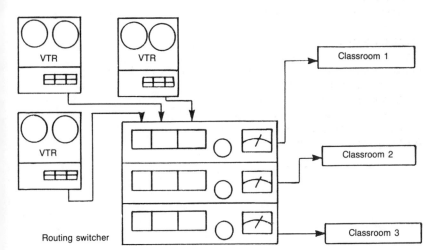

2

1. **On-air switcher.**

2. **Routing switcher.**

Production Switcher Flow Diagram

Take a look at the flow diagram of a switcher integrated into a simple video system. This switcher is very basic and a lot has been left out to keep things as simple as possible. As you can see, this system has three cameras and one film chain as video sources.

Now follow the outputs of each of the cameras and the film chain. In each case, the output goes to a distribution amplifier and from there one output goes to a monitor and another goes to the production switcher.

Switcher buses
You'll notice that at the switcher there are two rows of buttons, each row a duplicate of the other. These rows of buttons are called *buses.* They're what give you the ability to cut and dissolve between cameras.

Switcher outputs
The switcher has two outputs. The preview (PVW) output allows the director to see the next shot before it is used. The PGM output is what is intended to be recorded or transmitted. In this example, the final output is both recorded and transmitted.

The preview output is going to a preview monitor for the director's use. The program output is going to a distribution amplifier, and from there one output goes to the program monitor and another one goes to the on air-switcher where it is fed to the transmitter. The other two program DA outputs are going to VTRs where the show is recorded.

This is a simplified diagram of a very basic system, but if you study it, you'll get a good idea of how the basic components are integrated.

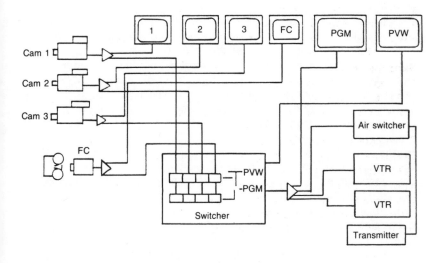

Switcher flow diagram.

Switcher Transitions and Special Effects

All production switchers are capable of at least three transitions: cuts, fades, and dissolves. *Cuts* are instantaneous changes from one picture to another. A *fade-in* is a transition that starts with a blank screen (black) that progressively grows brighter until the full picture appears at its normal intensity. A *fade-out* is the opposite, beginning with a full picture that decreases in intensity to a blank screen. A *dissolve* is much like a fade except that as one picture is fading out, another is fading in, so there is always a picture on the screen.

Cuts, fades and dissolves represent the meat and potatoes of television, as they make up most of the transitions used in dramas and comedies. All the other fancy effects that a switcher can do are often called the "bells and whistles." They're there for flash, sparkle, and pizzazz. If your content is solid, you don't need many bells and whistles, although the American public has come to accept them as part of the package. This is not to say that these special effects are valueless. Some can, and often do, add to the content of the program.

Wipes

Wipes are transitions between video sources that are marked by visible edges (sometimes the edge is diffused). Rather than one picture fading out as the other fades in as with a dissolve, in a wipe the new picture replaces the old one by means of a geometric form moving through the old picture. It might be a horizontal or vertical line moving across the picture, or it might be a star or circle coming from the center and expanding until it takes over the whole picture. The number of wipe patterns available seems unlimited. Some switchers come with 40 standard wipe patterns, with still others available as options. Wipes that aren't completed, thus leaving parts of two pictures visible, are called *split screens.*

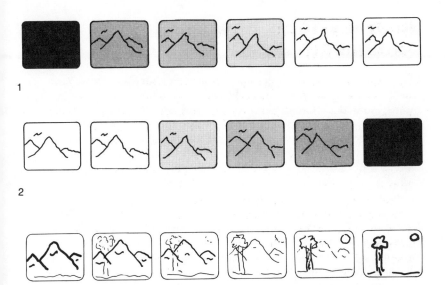

1

2

3

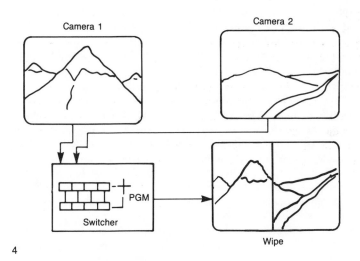

Camera 1

Camera 2

Switcher

PGM

Wipe

4

1. **Fade-in.**

2. **Fade-out.**

3. **Dissolve.**

4. **Wipe.**

Special Effects Keys — Luminance Keys

Keys are among the most common and useful special effects. Keys are essentially holes cut into a video picture that are filled with material from another source. Different types of keys rely on different information to determine the shape and nature of the holes and how they are filled.

Luminance keys are triggered by differences in brightness (contrast). The hole in the original video is cut according to contrast differences supplied by a programmed source. The pattern the system derives from this source is called the *key source* or *key signal*. The figure at right shows a common use of luminance keys.

Camera 2 is providing the key source — the bright star against the black background. The image camera 2 is shooting may come from a card of the black background and white star. The switcher receives the signal from camera 2 and reads the star as a pattern, the key source, based on the difference in brightness between the star and the background. The system cuts a hole in the shape of a star in the camera 1 video.

The hole can be filled in one of several ways. If the pattern that cuts the hole also fills the hole, it is a *self-fill key.* In other words, if the pattern is the white star on the black background, in a self-fill key the hole is filled with white. If the hole is filled by artificially created color from within the switcher itself, it is called a *matte key.* For example, using a matte key, the star-shaped hole in camera 1 could be filled with blue, even if the original key source was black and white. The hole can also be filled with a third source — video from a third camera, for example.

The greater the key source contrast between the desired pattern and its background, the more easily the key operates. This is particularly true when you are shooting a card as a key source, as opposed to using a programmed pattern. White on black is ideal, but for key sources that don't have as much contrast, there will always be a "clip" adjustment that will allow you to compensate. For example, if you are using a black card with a yellow star on it to operate a self-fill key, it will probably work — you'll get yellow color in your video hole — it just might not work as well as white on black.

Keys don't have to be geometric patterns. They can be (and often are) white lettering on a black background, titles, for instance. You certainly can key in other shapes, such as a white horse running against a dark background.

Camera 1

Program

Camera 2

Luminance key.

Special Effects Keys — Chroma Keys

Chroma keys are similar to luminance keys in that a hole is cut out of video, but unlike luminance keys, the triggering device is not contrast. It's a particular color in the subject video (the video receiving the special effect). In a chroma key, the system detects the chosen color in the subject video and wherever it sees that color, replaces it with information from another video source. In the first figure the talent is in front of a chroma key window. The window can be any color, but blue and sometimes green are most often used. The camera that shoots the chroma key window is called the *source camera.* Any other video source can supply the *fill* video, but in this example camera 2 has been used. Where the system sees the selected color (the window), it replaces that information with the corresponding information from the fill camera. Thus you get a rocket launch in the studio. If you were to make the entire background the chroma key window, then that background would be filled. The second figure illustrates this. You now see the entire area behind the talent filled with the launch because the entire background has become the key window.

The chroma key circuitry will try to fill in the source anywhere it sees the chosen color. It is for this reason that primary colors are used for the key window. If you chose yellow for your key window, for example, the system would not only lock up to yellow, but it would try to lock up to anything containing yellow's components, red and green. Thus, the chroma key would try to lock up to almost everything but the color blue.

Since people almost always appear in the chroma key source picture, red is not used for the key window very often. After all, there is quite a bit of red in the flesh tones of people. Although red chroma key windows with people in front of them have been used, it's not a common practice. Blue is the most commonly used color for chroma key windows, with green used occasionally.

It should be noted that talent wearing clothing the same color as the chroma key window will present problems, since it might be keyed out as well. Two other colors can create problems with chroma keys when they appear on the key source. Since white includes all colors of light, the chroma keyer may try to lock up to it, but this will usually appear as an incomplete key. Although black is the lack of colors, it too can create problems. Small patches of black, particularly shadows, may also be keyed over. When the system is scanning and it comes to a void of color (black), it may not be sure what to do so it keys over the area. Larger areas of black may not cause a problem because they're large enough for the system to determine exactly what they are.

Just as with the luminance key, the chroma key also has a clip control. In addition, there is a *hue* (chroma) control that allows you to choose the color to be keyed out, and a *gain* control that controls the strength of the source picture.

74

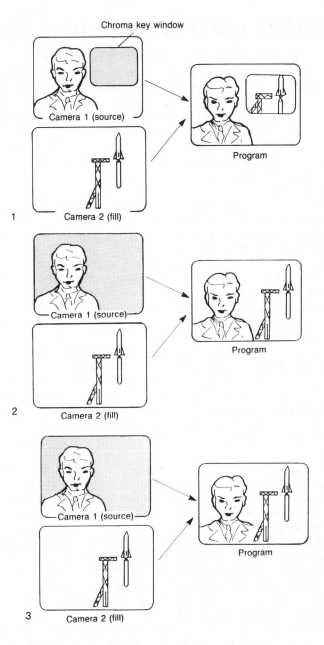

1. **Small chroma key window.**

2. **Full background chroma key window.**

3. **Talent wearing coat same color as key window.**

Digital Special Effects

Wipes, split screens, luminance, and chroma keys make up the majority of special effects on the typical production switcher. There is another group of special effects called *digital video effects.* While digital technology will be discussed in greater detail later, it is appropriate to describe here some basic digital effects. Digital effects units are normally added on to the switching system, although some of the newer switchers have them built in. Each make of digital effects unit will have its own repertoire of effects, often including the following.

Compressions

Compressions are effects that change the entire aspect of the picture: it's made longer, or higher, or the entire picture is compressed into a smaller area. A common use of this is called *chroma key tracking.* With this effect the entire fill video is compressed to fit into the chroma key window. In the first figure, the entire launch has been compressed to fit into the smaller chroma key window. Many stations avoid the chroma key process entirely by compressing what would be the fill video and inserting it into a predetermined space in the program video.

Pushes

Another common digital effect is a *push.* A push, as you might expect, is where one video source pushes another off the screen.

Flips

There are several types of *flips.* In a page flip the picture rotates around one edge of the screen as if you were turning the page of a book. Other flips can rotate around a central vertical or horizontal axis.

Rotations

Rotating cubes and spheres are two other digital effects, with video images composing the outer surfaces of these geometric shapes.

Other special effects

The array of possible digital effects is virtually limitless. Video images can be twisted, distorted and curled, and exploded into fragments and miraculously reassembled. In most situations, the limits on the imagination are only imposed by cost and time. Obviously, the more elaborate the effect, the greater the time and money necessary to produce it.

Many digital effects became practical and affordable due to development of programmable switchers. These switchers can hold in their memories a sequence or combination of special effects and then recall the arrangement on a command from the user.

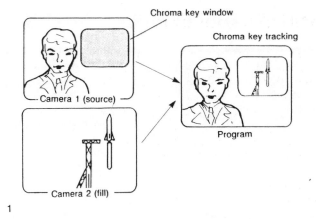

1

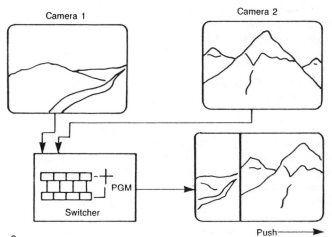

2

1. Chroma key tracking.

2. Pushes.

Videotape Recording Technology

Recorders

The development of the videotape recorder (VTR) was fundamental to the growth of modern communications. Videotape recording enables the practical recording and immediate or future playback of a high quality video image.

The basic theory behind the recording of video information is the same that allows us to record sound on tape. Just as with an audio cassette, video recording imprints magnetic signal patterns onto a specially prepared tape. Two of the key aspects of this process are the tape itself and the recording head.

Videotape

The foundation of videotape is a strong plastic ribbon. On the back of the tape is a slick surface that helps the tape move through a mechanical transport smoothly. On the front side of the tape are *metal oxides* mixed with a binding compound that secures the oxides to the tape. You might think of the metal oxides as being microscopic metal filings. If you pass a magnet close to these oxides, two things happen. First, their physical arrangement will be changed, and second, they will be left with their own, much weaker, magnetic field. The stronger the magnet that arranges these oxide particles, the stronger the induced magnetic field.

Recording heads

In order to record the desired information, a special electromagnet called a *recording head* has to be used. For video, the head is very small, made of very thin metal (about the thickness of a fingernail). The head is hollow, like a tube. It is often shaped something like a piece of bread. The curved surface is the one through which electrical information is exchanged with tape. Thin wire coiled around the other side of the head connects the head to the rest of the recorder.

In the second figure, the head has been enlarged many, many times. Although the figure shows the head by itself, in reality it would be mounted in a small nonmetallic enclosure. As the changing analog voltage (the composite video signal) from the camera electronics is processed and flows through the head, it causes corresponding changes in the electromagnetic field the video recording head produces. This, of course, leaves varied magnetic fields in the oxides on the tape. This is the basic recording process.

The playback process is just the reverse. The magnetically encoded tape is passed across a video head that has no signal flowing through it from another source (such as a video camera). The magnetic field on the tape induces a signal into the head corresponding to the varied magnetic fields on the tape. Thus you reproduce the same analog signal off the tape that was induced onto the tape from the camera source. Since the tape itself wasn't physically changed during the playback process, the signal information is still there for playback again.

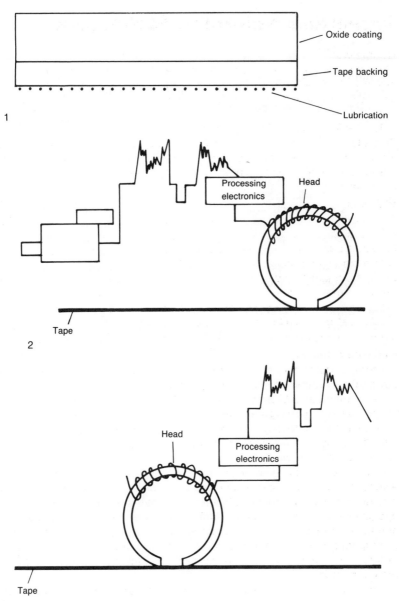

1. **Cross-section of magnetic tape.**

2. **Recording on tape.**

3. **Videotape playback.**

Video Recording Standards and Formats (1)

Audio versus video recording

The recording process is essentially the same for both audio and video recording, but there is about 300 times more information in the video signal as there is in an audio signal. There is only so much information you can fit into a given piece of tape. As a result, there's just not enough room for all the video information using audio recording technology.

There are thus several significant differences between video and audio recording technology that reflect video's need for greater information capacity. First of all, standard audio tape is a quarter inch wide. Early video tape was much larger (two inches), although with new technology it is generally one half to one inch wide. Second, standard audio tape speed is 7 ½ inches per second (ips), while video runs at 15 ips. Finally, audio heads are stationary. Video uses several smaller heads mounted on a rotating disc. All of these characteristics provide video with the extra information capacity it requires.

Quadraplex video recording

The first successful broadcast quality system was called *quad* or *quadraplex.* In this system, four video heads are mounted equidistant from each other on a rotating disc called a *headwheel.* The disc rotates perpendicular to the path of the tape. It takes one pass of each head or a total of four passes to lay down one field of video information. If you were to stop the tape during playback and keep the headwheel rotating, what would happen? Wouldn't you keep playing back the same quarter of one field over and over again? Well, that's exactly what would happen. Also, if you simply slow down the tape speed, you won't get 60 complete fields a second. As a result, freeze frames are not possible with quad machines, nor is slow motion. In addition, quad machines are large, heavy, and expensive. For these reasons they are quickly being replaced by smaller, lighter, less expensive, and better quality machines. However, quad machines are very reliable. Although new quad videotape recorders haven't been manufactured in a few years, the machines already at stations will probably continue to be used for a number of years.

80

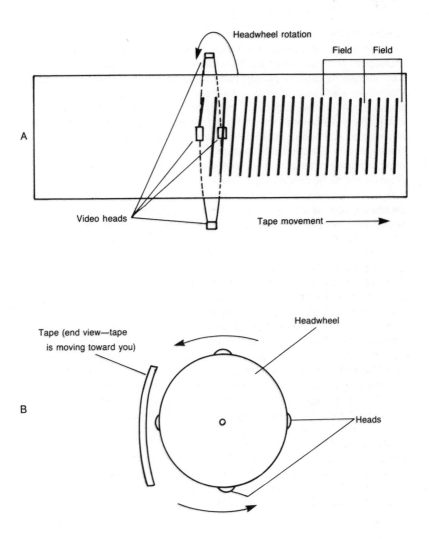

Quadraplex format. (A) View from back side of tape; (B) view from side of head wheel.

Video Recording Standards and Formats (2)

Helical video recording

Helical tape machines have become the new standard of the broadcast industry. There are at least a half dozen different formats of helical video-tape machines that might be found in professional broadcast operations. While none will interchange with the others, they all have some things in common. While the number of heads mounted on the headwheel will vary with different formats, unlike the perpendicular positioning of quad head-wheels, helical headwheels are mounted at an angle and rotate in almost the opposite direction of the tape. The magnetic heads lay down long slanting video tracks on the tape. Each of these tracks will contain one field of video information: video lines, blanking, and sync information. Therefore, if you were to slow down or stop the tape with the headwheel spinning, you would still get 60 fields of information a second, so slow motion and freeze frames are possible with helical machines. However, because of some other problems that will be discussed later, these are not broadcast-quality slow motion or freeze frames.

SMPTE Type C and U-Matic formats

The two most common helical broadcast formats are the *SMPTE* (Society of Motion Picture and Television Engineers) *Type C* and the *U-Matic®*. Type C, which uses tape that is one inch wide, is smaller, lighter, less ex-pensive, and produces better pictures than the older quad format does. It has become the new standard for the American broadcast industry. The U-Matic format, which uses ¾-inch-wide tape, is much smaller and lighter, and is far less expensive than either the two-inch quad or Type C machine. However, the current U-Matic does not produce as good a picture as the other two machines. Therefore, its uses in broadcasting are more limited at the moment.

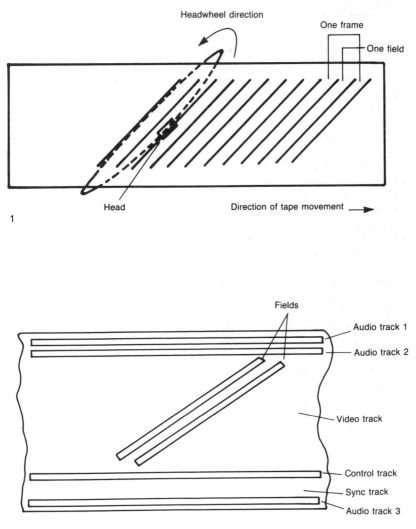

Headwheel direction

One frame

One field

Head

Direction of tape movement →

1

Fields

Audio track 1

Audio track 2

Video track

Control track

Sync track

Audio track 3

2

1. Helical format.

2. Various videotape tracks (SMPTE Type C).

83

Video Recording Standards and Formats (3)

Other broadcast quality formats

There are also broadcast-quality machines that use half- and ¼-inch tape. These machines are extremely small, light, and electronically complex, but they produce excellent pictures. The two most commonly found formats in half inch are Sony Corporation's Betacam and Panasonic's M-II. Both of these formats use what looks like a normal consumer videotape cassette. However, the professional formats record at a higher speed and use a special high-quality tape in the cassettes.

Sound and control tracks

So far only the video information recorded on tape has been discussed, but there is a lot of other information needed on the tape. Sound, for example, also has to be recorded. All of the nonvideo information is recorded using stationary heads much like the heads used in standard audio tape recorders. The number, type, and placement of these other tracks will vary with the specific format of machine being used.

One of the most important of these other tracks, and one that is common to all formats, is the control track. During the recording process, the vertical sync pulses are recorded on the control track. The control track thus helps stabilize the tape's playback speed. You know that the vertical sync pulses are laid down at a rate of 60 pulses a second (one for each field). The regularity of these pulses makes the control track important for other reasons that will be discussed later.

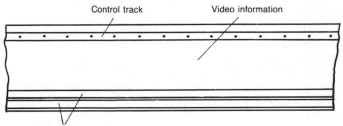

Control track   Video information

Audio (sound) tracks

A section of ¾ inch U-Matic format tape showing control and sound tracks.

Time Base Error

The *tape transport system* is a very important and complex mechanism that pulls the tape across the tape heads. It is also the source of a major problem in videotape recording. The problem is that it's impossible to build such a system that operates at a truly constant speed, and that means that it is impossible for the machine to play back tape at precisely the same speed at which it was recorded. You can come pretty close, but in some cases that's not good enough.

If you review the section on sync generators, you'll recall that the video signal is composed of some very precise bits of information. Since all this video and sync information is coming from the sync generator and camera, it is being recorded very accurately. But since the VTR can't play back the tape at exactly the same speed at which it was recorded, the playback information won't be as precise as what was recorded. This inability of a VTR to play back at exactly the same speed is called *time base error*. Time base error is measured in lines. It takes the electron gun 63.5 μsec to scan a video line. If the playback signal is 63.5 μsec off from where it should be, there is one line of error.

Time base error can thus be an integral part of videotape recording. If the amount of error is small or can be corrected, the resulting picture problems will be minimal. However, a large error or, in some cases, any error at all can produce terrible jitters, jumps, and roles in the picture.

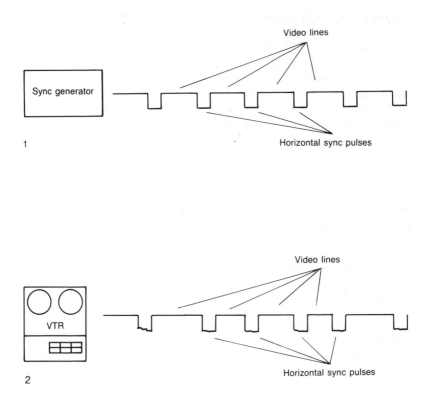

Video lines

Sync generator

Horizontal sync pulses

1

Video lines

VTR

Horizontal sync pulses

2

1. Sync generator puts out very stable signals with each video line being recorded at exactly the same length and the horizontal sync pulses reproduced at the same intervals. This precise signal is then recorded on videotape.

2. Because the videotape machine can't play back at an absolutely precise speed, the video lines vary somewhat in length and the horizontal sync pulses are reproduced at varying intervals. Note the degradation of the sync pulses, which comes from generation drop and not from time base error. This problem and a solution are discussed on pages 96 and 114.

External Causes of Time Base Error

A VTR that spends all of its time in the studio will have relatively little time base error, assuming the equipment is in good condition and properly used. But VTRs that are used in the field have an additional chance for time base error. Time base error can result from any change in the actual composition of the tape, generally due to variations in air temperature, humidity, recorder position, and recorder movement. When a portable machine is used in the field it is subjected to such constantly changing conditions. It might be in the bright sun one minute, and in the cool shade of a tree the next. You might be shooting at the fog-shrouded seashore in the morning, and in the hot, dry desert later that afternoon. Changes in temperature and humidity will cause the tape to expand and contract. How might this result in time base error?

Assume that you're shooting outside on a hot, sunny day. This will cause the tape to expand. After a hot afternoon of work, you return to the nice air conditioned studio to edit the piece. The cool temperature of the studio causes the tape to contract. Well, even if the VTR could play back the tape at exactly the same speed it was recorded at, there would still be a problem. Since the tape has contracted, the video tracks have also contracted, so it will take a little less time for the head to scan them. Additional time base error has been created.

Gyroscopic time base error

Gyroscopic time base error refers to the creation of time base problems specifically by changes in recorder position and movement. It's a very common occurrence given today's highly portable video equipment. Let's say that you've got one of those great camcorders that combine the camera and VTR in one unit. You have it up on your shoulder for some great shots and then you swing around to follow the action. *Gyroscopic time base error* has been created. Gyroscopic error occurs because the spinning head drum in the VTR acts like a gyroscope. If you have ever played with a toy gyroscope or a spinning top, you know that if you try to move the toy against the plane of its rotation you will feel resistance. This is what happens inside the VTR. As the VTR is moved against the plane of rotation of the head drum, the spinning drum resists and slows down a little. When it slows down, the information will not be recorded at the proper speed. This is what gyroscopic time base error is.

How much time base error you might expect on tape playback depends on all of these factors. It could be less than a line in a studio machine, or add up to 20 or 30 lines or more on a field recorder!

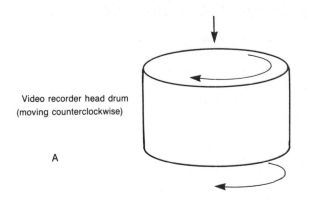

Video recorder head drum
(moving counterclockwise)

A

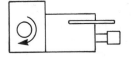

B

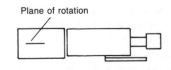

Plane of rotation

C

D

Gyroscopic time base error. (A) Forces that move against the head drum's plane of rotation will cause gyroscopic error; (B) side view of camcorder showing head drum and rotation; (C) top view of the same camera showing the head drum's plane of rotation; (D) gyroscopic error is caused when the head drum's plane of rotation is changed.

A time base corrector stabilizes the video signal, allowing precise playback.

Time Base Error Correction

Quad

Quad tape machines have relatively little error and the circuitry to correct that error is fairly simple. Besides, they're much too big to pack around in the field. As a result, quad machines don't present much of a problem. Helical machines, on the other hand, can have a great deal of error and until fairly recently there was no practical way to correct it.

Helical

Time base error becomes a real problem when you try to integrate video-tape material into a production. If you just want to play back the tape, no problem, but if you want to fade, dissolve, wipe, split screen, or key using taped material, forget it! You've got an imprecise playback signal trying to match up with the very precise, sync generator–controlled video system. If you try to do any of the above-mentioned effects with tape, the picture will jump, jitter, roll, or tear. In short, it will look terrible. This can be corrected with a *time base corrector,* which can cost as much as some VTRs, although the prices have come down dramatically in the last few years. A time base corrector will take the unstable signal coming out of the VTR and stabilize it. Exactly how they work will be discussed later.

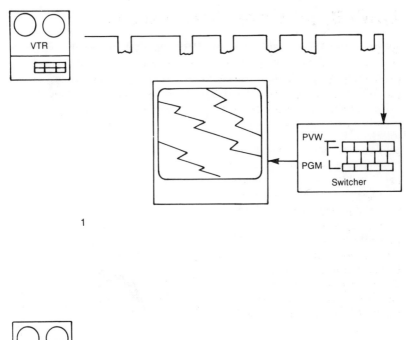

1

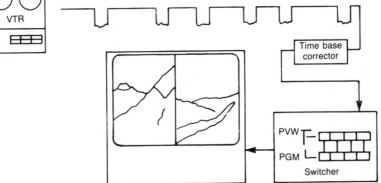

2

1. Integrating helical VTR video without time base correction.

2. Integrating helical VTR video with time base correction.

Several system features can help ensure stable video playback.

VTR Lockup (1)

The more precise we can make the VTR's playback speed, the less time base error will be created. When the machine gets up to full speed and everything is as stable as it is going to get, we say the machine is "locked up." There are several degrees of lockup and each additional step adds a little more stability.

Capstan lock
The first degree of lockup is called *capstan lock,* suitable mainly for home videotape recorders. Capstan lock is not a very stable state. The machine essentially relies on the stability of the power source for a constant base. If the power coming out of the wall varies, so does your tape speed, because the speed control circuitry is very simple.

Vertical lock (capstan servo)
The next level is called *vertical lock* or *capstan servo.* This is the minimum degree of lock up needed to do videotape editing. Capstan servo machines are considerably more complex than capstan lock machines and make use of the control track (remember it?) and incoming sync from the sync generator. If you'll recall, vertical interval pulses are laid down on the control track and, since they come from the sync generator, they are laid down at very precise intervals. A capstan servo machine is also hooked up to the sync generator and has a special circuit to compare the number of incoming vertical sync pulses from the sync generator with the number of pulses being played back from the control track. Since both pulses ultimately originate from the same source (remember that the control track pulses originally came from the sync generator as vertical blanking pulses), there should be the same number of pulses in the same amount of time. Since the vertical sync pulses come from the sync generator at precise intervals, they act as a clock. If there are 60 pulses from the sync generator and only 55 from the control track, then the tape is moving too slowly and the capstan servo is signaled to increase the playback speed. But if there are 60 pulses from the sync generator and 63 from the control track, then the tape is moving too fast and the capstan servo is signaled to slow it down.

The big advantage of vertical lock is that a vertical interval switcher will be able to switch to tape without a breakup. However, time base error will still be a problem and dissolves, wipes, and other special effects with tape will not be possible.

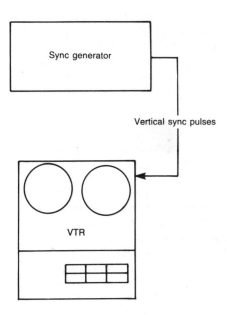

Sync generator

Vertical sync pulses

VTR

The capstan servo machine compares the incoming vertical interval sync pulses with those recorded on the control track to more accurately control the tape's playback speed.

VTR Lockup (2)

Frame lock

The next level of lock up is called *frame lock* and follows the same rule of comparing information from the sync generator with playback information to help regulate tape playback speed. With vertical lock, the system was just trying to match one vertical sync pulse from the control track for each vertical sync pulse coming in from the sync generator. There was no attempt to match an even field pulse with an even field pulse or an odd field pulse with an odd field pulse. That's what the frame lock circuitry does. It determines whether the vertical sync pulse coming from the sync generator is for an odd or an even field. Then it speeds up or slows down the tape machine until the pulses off the control track match: odd for odd and even for even. This makes the tape playback speed just a little more precise.

Horizontal lock

We now have the fields matched, but each field has 262.5 lines, and each line has a horizontal sync pulse. This leads to the next level of lock up, *horizontal lock.* The horizontal lock circuitry compares the number of incoming horizontal sync pulses with the number of played back horizontal sync pulses. The VTR then speeds up or slows down the tape in an attempt to match the two horizontal pulses.

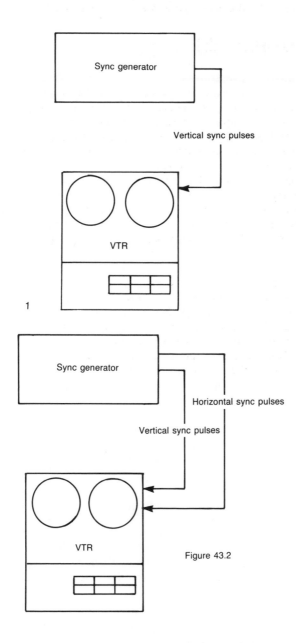

Figure 43.2

1. A frame lock machine will try to match the tape playback of an odd field (lines 1, 3, 5, 7, 9 . . .) to the sync generator's odd field and an even field to an even field.

2. The horizontal lock machine tries to match every horizontal sync pulse played back to a horizontal sync pulse from the sync generator.

Editing Videotape

Physical cutting and splicing

In the early days of videotape, editing was a difficult and time-consuming process. A liquid solution had to be put on the tape to make the control track oxide patterns visible. The tape had to be viewed under magnification and then it was physically cut. Finally, the ends were spliced together with a special tape. On top of all this, the roughness that the splice created in the tape's surface often damaged or even destroyed the tape head. Needless to say, editing tape was done only when there was no other alternative.

Electronic editing

Today, however, video editing is done electronically. Information that should be sequential may be out of order on one tape, or it may be distributed between several different tapes. During editing, the material is recorded electronically in the proper sequence onto another tape. The process of copying information from one tape to another is called *dubbing*. Whenever we dub a tape, we drop a generation. The original tape that video is recorded onto is called the *master* or *first-generation* tape. If we make a copy of that tape it is a *second-generation* tape. If a dub is made of the second-generation tape, the resulting tape becomes a third-generation tape, and so on. Every generation drop also entails a loss of quality. The signal has to go through the entire electronics of a tape recorder, be recorded onto the tape, taken off the tape, and run to another tape recorder. Every step of this process adds additional noise to the signal and diminishes the original quality of the signal. A top-rate VTR, such as an SMPTE Type C machine, can go several generations before this quality loss becomes visible to the human eye. However, a low-quality machine, such as a home VTR, will show a very visible quality loss on a second generation tape.

Editing allows you to shoot a video production in the most efficient manner, which often means shooting scenes out of sequence from the script. These disconnected scenes then can be rearranged and put together in sequence in the editing suite. The second figure shows the master tape, with the shots in nonsequential order. Through the editing process, the shots or scenes are arranged into their proper sequence.

1. Tape generations.

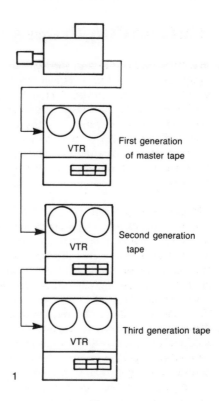

First generation of master tape

Second generation tape

Third generation tape

1

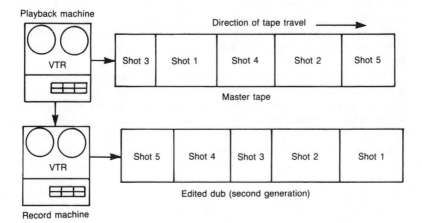

Playback machine

Direction of tape travel ⟶

| Shot 3 | Shot 1 | Shot 4 | Shot 2 | Shot 5 |

Master tape

| Shot 5 | Shot 4 | Shot 3 | Shot 2 | Shot 1 |

Edited dub (second generation)

Record machine

2

2. Tape editing.

The basis of editing is putting two shots together.

The Editing Process

It will be helpful to step through several edits made on a basic editing system just to become familiar with some concepts in the process. You begin with the master tape (containing the shots in nonsequential order), a blank tape, and the VTRs on which to run them; the playback VTR, which will run the master tape; and the record VTR, by which edits will be recorded on the blank tape. First, you find the beginning and end points of the first shot, also called *edit points,* and stop the master tape in the playback machine at the beginning of that shot. Preparing a tape for editing in this manner is called *cueing up* the tape. The first shot will be recorded onto the blank tape.

The motors in each VTR must be at full speed and have achieved their full level of lock up (see pages 92–95) before accurate recording can begin. Most modern editing machines take five to ten seconds to reach full speed and lock up. Thus, rather than beginning the recording process at the first edit point, you must start the VTRs approximately five to ten seconds before this point. This is called *prerolling* the VTRs. You stop the machines when the edit has fully been recorded onto the blank tape.

You then locate on the master tape the edit points for the second shot. The beginning of the second edit is cued up on the playback VTR, and the tape in the record VTR is stopped at the end of the first edit. Both machines are prerolled and then started in the *playback* mode. At the precise instant the playback and record VTRs reach, respectively, the beginning of the second edit and the end of the first, the record VTR is switched into the record mode. After the second edit is fully recorded, both machines are stopped and the entire process can be reviewed. You should be viewing shot one with a sharp, clean cut to shot two at the right moment.

You go through the same process for the rest of the shots in your show. This is how you build scenes into sequences, and from sequences into final shows.

The process just described concerns by far the most basic edit possible and uses the most rudimentary equipment. The more advanced and precise editing systems and methods are, the more complex the means by which video information is organized, accessed, and transferred into a sequence.

In a very basic system such as that described, the operator may unscientifically gauge when to make cuts and punch the specific button where he or she wants material to be copied. In advanced systems, such tasks are electronically programmed into a computer and precisely executed according to extremely accurate timing systems.

On longer videos especially, the editing process can be quite complicated. A crucial part of most editing tasks is the *editing script.* An editing script is a list of the shots you want to use in the order you want to use them. The script will include the beginning and ending points of each shot. In very basic systems these points may be specific visual or aural refer-

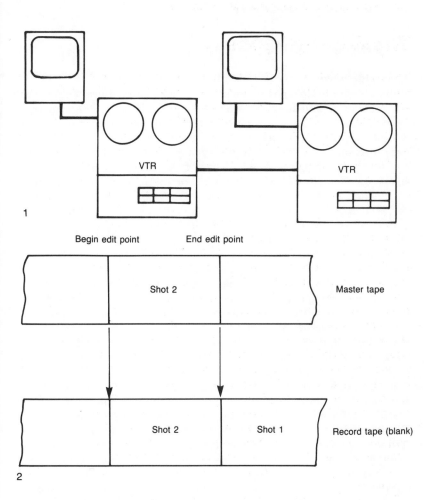

Begin edit point   End edit point

VTR   VTR

1

Shot 2   Master tape

Shot 2   Shot 1   Record tape (blank)

2

1. **Manual editing.**

2. **A typical video edit. A segment of tape is recorded from a master tape to its proper position on the record tape.**

ences and you may need to manually time the length of each edit. Advanced systems have special coding systems by which you can designate the edit points.

Although creating an edit script may require a good deal of time, having a good script before you begin to make the edits will save you a great deal of effort. Ideally, most of your editing decisions will have been made ahead of time by reviewing the tapes so that you have less to worry about when you sit down to actually make the edits.

A shot can be sequentially placed in several ways.

Types of Edits

Assemble edits

There are two basic types of edits: *assemble edits* and *insert edits.* With assemble edits, you start with a blank tape to record on. When you make the edits onto that tape, everything is recorded from the master tape: audio, video, and control track. Since the control track is coming off the playback machine rather than from the sync generator, the interval between pulses may be off (because of time base error), and the pulses themselves will be degraded (because of the generation drop). Because of these problems, when the edit is played back the images may break up on the monitor. The advantage of assemble editing is that it takes no advance preparation of the tape, but there may be problems in getting good clean edits.

Insert edits

Insert edits, on the other hand, require some preparation of the recording tape before editing begins. In order to do insert edits, a control track must be laid down first. This is usually done by placing a blank tape into a VTR, punching up black on the switcher, and putting the machine in record for the entire length of the tape. This will record a black video on the video tracks, and lay down the control track pulses from the sync generator.

Now you are ready to do insert edits, and insert edits give you a couple of advantages. First, you can edit using the procedure outlined above, but since the machine is using good, clean , consistent control track pulses, the edits are less likely to break up. Second, you can insert new material over old material. Assume that you had tape of a person talking about a terrific new invention. The audio information is interesting, but the video of a talking head is boring. So you take interesting video that you have of the invention (called *B-roll* material) and insert just that video over the boring talking head. Now, you have the interesting audio information as well as a visually interesting picture to look at. You could also insert new audio without changing the video you laid down previously. You can go either way. But the key thing to remember is that to do insert editing, the control track must be laid down on the tape you intend to edit to.

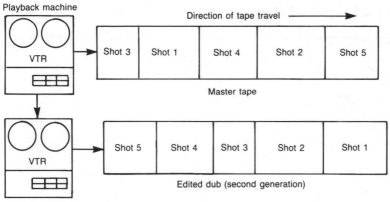

Playback machine

Direction of tape travel ⟶

| Shot 3 | Shot 1 | Shot 4 | Shot 2 | Shot 5 |

Master tape

VTR

| Shot 5 | Shot 4 | Shot 3 | Shot 2 | Shot 1 |

Edited dub (second generation)

Record machine

1

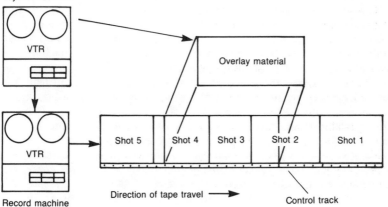

Playback machine

VTR

Overlay material

| Shot 5 | Shot 4 | Shot 3 | Shot 2 | Shot 1 |

VTR

Record machine

Direction of tape travel ⟶   Control track

2

1. **Assemble edits.**

2. **Insert edits.**

To be edited, video material must have a structure.

Editing Methods

There are three basic methods of controlling and editing videotape: manual, control track, and SMPTE time code.

Manual editing

Manual editing was described in "The Editing Process." You select and make edits based on visual or aural edit points. If, after prerolling both playback and record VTRs and placing them in playback mode, you decide the edit points are not coinciding as you'd like, you can abort the edit by not switching the record VTR to record mode. You can then change the amount of preroll on one or both tapes and try again. This is obviously a rather inaccurate and time consuming way to edit, but sometimes it's the only choice.

Control track counters

A better method is to use a *control track counter editing system.* As noted earlier in "Video Recording Standards and Formats (3)," vertical sync pulses are laid down regularly on the control track to stabilize the tape's playback speed. A control track counter editing system uses these pulses as reference points to simulate and perform edits.

The control track counter is hooked up between the VTRs, although some systems have a counter built in. This editing process begins much as the manual method. You select edit points on the playback and record machines. However, the editing system usually prerolls the machines for you and offers a *preview* option. If you select this mode, the system prerolls the machines and starts them, displaying first the output from the record machine (the previous shot) and then, at the correct edit point, switching to the playback machine's output (the current shot). This simulates the edit without actually making it. If the edit was good, you can have the system actually make the edit that was previewed.

This sounds like a pretty incredible machine, but it operates on a very simple principle. The heart of this editing system, the *edit controller,* merely counts control track pulses. So once the edit point is chosen, all it does is count pulses. When the controller backs up the machines for preroll, it backs both machines the same number of pulses, then rolls forward, counting the pulses, until it gets to the original edit points, when it makes the edit.

The controller ends the edit in the same way. Based on the end points that you predetermined, the controller measures the duration of the edit by counting the number of pulses. When you finally initiate the edit, the controller can recount the pulses and properly end the edit.

Since the controller is counting control track pulses, there had better be control track laid down in the areas the controller engages. This is particularly important in assemble editing when you are adding a shot. The control track pulses at the very end of an edit (in the last few frames) may

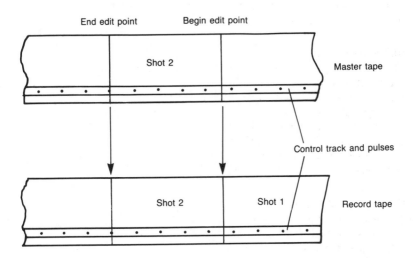

An edit made using a control track system. The system determines the length of the edit by counting control track pulses.

not be of good quality because the controller has already begun to end the edit. If your end point is exactly where you want the edit to end, the next edit's in-point may be either in an area of poor pulses or where there are no pulses at all. Therefore, when assemble editing, always let a scene continue longer than you need to; your next edit still can begin where you want it and you will have good pulses for the in-point.

Control track editing certainly is a vast improvement over manual editing, but there are still problems. The pulse counts sometimes may be unreliable as a means of measuring videotape material, since this isn't their original purpose. Control track editing is fine for some applications, but not accurate enough for many broadcast purposes.

In addition, since you can't fully rely on control track to measure video material, you must keep a very careful editing script, recording audio and visual references for the edit points and manually timing each shot. Even if you carefully log such information, finding the shots you want to use can be very time consuming.

103

SMPTE Time Code Editing

A much better (and more expensive) method of structuring and editing video material is *SMPTE time code.* SMPTE time code uses one of the audio tracks (see diagram on page 85) or a special *address track* that many formats have to lay down a specific code number for each frame of video information on the tape. This code consists of a time so that, for example, 00:27:14:03 would be read as 0 hours, 27 minutes, 14 seconds, and 3 frames. This system works well partly because there are 30 video frames per second.

Since each tape would normally start at 00:00:00:00, the above address would be almost half way into an hour tape. The address is permanent. If the tape is taken off the machine and stored for a few years, 00:27:14:03 is going to be at exactly the same place on the tape coming out of the vault as it was when the tape went into the vault. Since most professional tapes are no more than an hour long, the hours column of the time code is usually used to identify the tape. For example, a time code of 05:36:56:24 would normally be, 36 minutes, 56 seconds, and 24 frames into tape number 5.

The first advantage of SMPTE time code is that it makes putting together an editing script much easier and faster. With SMPTE time code, when you find the beginning or ending edit point, you simply write down the time code number. You don't have to write down a description of the aural or visual cues that indicate the edit points. Thus, it's much faster to go through a tape and put together an editing script. This is particularly important if the video is very long and/or complex, such as a television program.

Editing by SMPTE time code is much more expensive, but it is also much more precise and reliable. For this reason, and the fact that the code enables you to more easily organize and manipulate large amounts of material, most broadcast tapes are edited using time code.

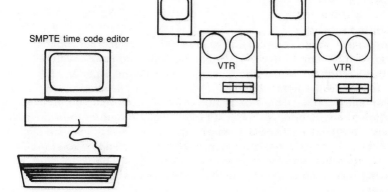

SMPTE time code editor

SMPTE time code editor.

Off-Line and On-Line Editing

SMPTE time code also makes on-line and off-line editing practical and economical. A top-of-the-line, full-blown Type C one-inch editing suite (the system is so large it requires a room) can easily cost over $1 million to build, while an editing suite using a smaller format, like ¾-inch U-Matic, might cost less than a quarter of that. The rental fees of such suites will reflect the difference in costs. So it's going to cost a lot more (five or six times as much) to do all your work in the one-inch suite than if you could do most of your work in the ¾-inch suite and spend only a little time in the one-inch suite. This is what on-line and off-line editing lets you do. To begin with, when you first record your show on one-inch tape, you also record it on ¾-inch tape and lay down the exact same time codes on the two tapes.

Off-line editing

During this phase you put your one-inch tape in a storage vault and take your ¾-inch tape to the U-Matic editing suite and start developing your editing script by viewing the tape and selecting the order and edit points of the video segments you wish to join. As you develop the script, you also edit the tape. This will be our work print. Developing your editing script and work print is the most time-consuming job of the editing process. You don't really care about the quality of the ¾-inch tape since no one else is going to see it. Its primary purpose is to allow you to see how the edits fit together so that you can finalize your edit script. Putting together your editing script and your small-format work print that no one else is going to see is called *off-line editing.*

On-line editing

Putting together the large-format final tape is called *on-line editing.* After you finish the work print and the editing script, you retrieve your one-inch tape from the vault and go to the one-inch editing suite. The final, high-quality, one-inch tape is easy to put together because during off-line editing you determined the specific end points and order of every edit. Since the time code numbers were the same for the ¾-inch and one-inch tapes, you can simply use your final editing script numbers to edit on-line.

Off-line editing.

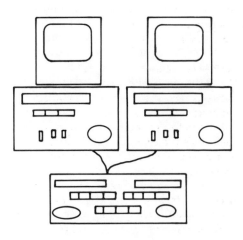

Computers help coordinate the editing process.

Editing by Computer

The real beauty of SMPTE time code editing is that special edit controller computers can do most of the work for you. The most sophisticated of these machines can control several tape machines and a programmable switcher at the same time. With more than one playback machine (provided they're properly time base corrected), you can dissolve, wipe, key, and do other effects between machines. All of this can be entered into the edit controller and it will do everything, cue and preroll the machines, command the switcher to do special effects between the machines, and have the record machine do the actual edits. All you have to do is program the computer properly by entering the editing script containing all the scenes in sequence designated by SMPTE time code.

All of this sounds incredible, and it is, but you have to keep one thing in mind. The computer can very accurately carry out your commands, but it takes the creativity and organization of a person to create, develop, and refine the show's idea and execution.

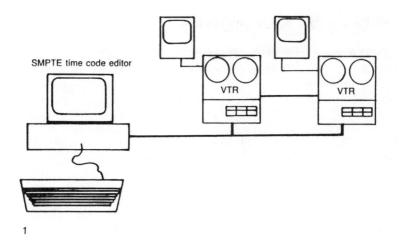

SMPTE time code editor

VTR

VTR

1

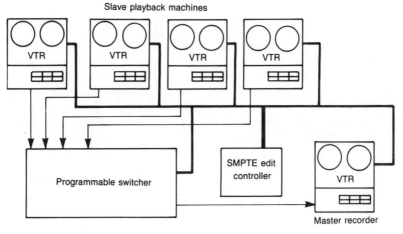

Slave playback machines

VTR

VTR

VTR

VTR

Programmable switcher

SMPTE edit controller

VTR

Master recorder

2

1. On-line editing.

2. SMPTE time code editor running entire editing suite.

Time Base Correctors

Review of time base error problem

In an earlier section you found that it was impossible to make a VTR that would play back tape at the exact same speed at which it recorded, and thus time base error is induced into the playback signal. In addition, changes in the environment can create error and gyroscopic errors can result from the movement of the machine. All of this time base error can create such an unstable signal coming out of the VTR that it can't be integrated into a video production through a switcher.

What a time base corrector does

Total time base error in a typical video can amount to more than 20 or 30 lines and needs to be brought down to the range of 5 to 10 nsec. A modern time base corrector (TBC) can do this job for us. Unstable video goes from the VTR into the TBC and stable video comes out of the TBC, which is then fed into the system.

How a TBC works

Of course, things are considerably more complex than this. The biggest problem is that we need a very large memory to store the video picture in so that it can be fed out at an even, synchronized rate. Memories that will hold a lot of analog video are very expensive and difficult to make. However, computer memories are relatively inexpensive and common, so if we could get our analog video into computer form, we could make good use of those memories. That's just what happens. But the only thing that computers can deal with is numbers. Even if you're typing on a computer keyboard, what is really happening is that the letters on the keyboard are being changed to numbers so the computer can use them. In the time base corrector, this process of changing video into *digital* information takes place in the *A-to-D* (analog to digital) *converter.* From there the information goes to a large computer-like memory.

110

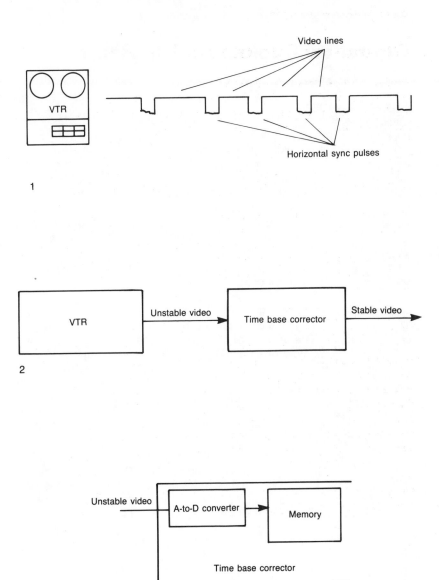

1. Because the videotape machine can't play back at an absolutely precise speed, the video lines vary somewhat in length and the horizontal sync pulses fall at varying intervals.

2. What a TBC does.

3. A-to-D converter in a TBC.

Digital technology can solve many video problems.

Digital Technology — The Heart of a TBC

A-to-D Conversion

The process of converting information from analog to digital is fairly complex, but it is important that you understand as much of it as you can.

You know that the video signal is up to .7 V, and that the stronger the voltage (.7) the brighter the glow on the CRT; the weaker the voltage (.00), the darker the CRT. If we picked a point right in the middle of that range (.35 V), we could probably figure out that that would be in the middle of the gray range. This is sort of what the A-to-D converter does. The brightness range is divided into 256 possible levels with .00 V equaling the 0 level and .7 V equaling level 255. Midway through the voltage range (.35 V) would also be midway through the number range (127). As a result each of the 256 numbers represents a separate and distinct voltage and brightness (remember, these voltages have been rounded off).

Sampling and quantizing

The process of grabbing a piece of the video information and holding it is called *sampling.* The process of changing that sample into a number is called *quantizing.* How often a sample is taken is very important. If we sampled the video only a few times a line, we wouldn't get a very accurate picture of what was happening. How often the video is sampled will differ between various makes of equipment, but it will always be a factor of the color burst frequency (3,579,545 Hz). Normally, either three times (around 10.7 M) or four times (around 14.3 M) the color burst frequency is used. In either case, each line of video will be sampled and quantized several hundred times. If you think of the brightness levels along one axis and the sample points along the other axis, you can turn the waveform display into a point-by-point graph that is quite representative. That has been done in the third figure. Imagine how much more accurate it would be with several hundred sampling points instead of a few.

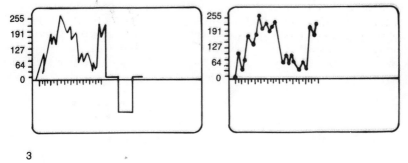

1

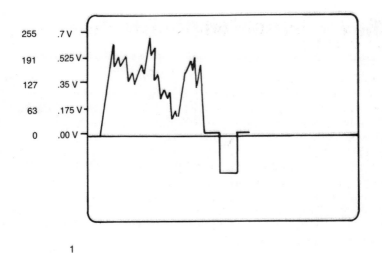

2

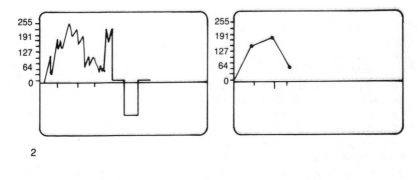

3

1. **Breaking the video signal down to numbers.**

2. **Sampling only a few times a line.**

3. **Sampling several times a line.**

Your result is a corrected analog video signal.

The Synchronized Signal

Horizontal sync as a clock
This is the way, then, that a continuously varying voltage goes into the A to D converter and comes out as a stream of numbers. But a method needs to be devised to release this digital video information in sync with the rest of the system. As you might suspect, we go back to the sync generator, which is hooked up to the TBC. The horizontal sync pulses from the sync generator act as a clock. Each time a horizontal sync pulse reaches a special gate circuit, it lets one line of digital video information out of the memory.

D-to-A conversion
This digital video can't be integrated with analog video, so it has to go through another stage. The *D-to-A* (digital to analog) *converter* changes the digital information back to analog, the reverse of what the A-to-D converter did. As a result, corrected analog video comes out of the TBC.

Video proc amp
In addition, most TBCs have an internal *proc amp* (video processing amplifier). As previously noted, the quality of the sync itself may have deteriorated due to factors such as copying a tape. However, proc amps have their own sync generators, so that when the distorted sync comes off the tape, the proc amp strips it away and inserts new clean sync in its place. Proc amps also enable you to adjust some of the video parameters such as brightness, chroma intensity, phase, and pedestal (black level).

Window of correction
Each TBC will have what is called a *window of correction.* As you might suspect, this tells you the maximum amount of time base error it can correct. A TBC with a four-line window might be fine for VTRs and tapes that never leave the climate-controlled confines of a studio, but it would be almost useless with tape that is shot out in the field. A TBC with a 32-line window will cost a good deal more money, but it should be able to handle anything shot in the studio or the field.

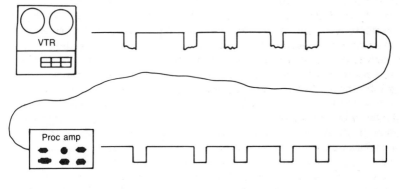

1

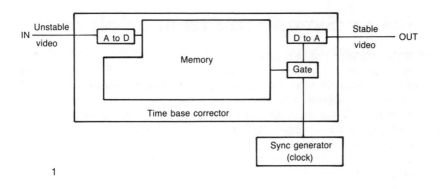

2

1. Time base corrector.

2. Proc amps strip off old distorted sync and insert new, clean sync.

115

Larger Sync Problems and Solutions

Nonsynchronous sources

TBCs are great for correcting the relatively minor errors found on tape, but there are some video sources that are totally out of sync with the studio. For example, when the networks do a football game, do you think there's a cable going from the sync generator on the ground all the way up to the blimp that's getting those dramatic aerial shots? Of course not! Since there is no incoming sync for the blimp camera, it has to be on its own system and it is totally out of sync with the cameras on the ground. We say the blimp is a *nonsynchronous* video source. Or, what about the local TV station? Each station will have its own sync generator, but the video coming into the station from a network or a satellite feed will be on a different sync generator and thus be a nonsynchronous source. If the local station wants to dissolve from the network football game to a local commercial, it could lose sync and the picture will break up.

Frame store synchronizer

The solution to problems posed by nonsynchronous sources is the *frame store synchronizer* (FSS). While the FSS operates a little differently than a TBC, you can think of it as a TBC with a huge memory. It can hold more than a frame of video information in its memory. Like a TBC, an FSS converts lines of analog video signal into digital form and stabilizes them. However, the vertical sync rather than the horizontal sync acts as the FSS gate release signal. When it gets a vertical sync pulse from the house sync generator, the FSS feeds out a field of video information. As a result, any nonsynchronous video source can be locked up to and integrated with the system you're using. Frame store synchronizers are pretty expensive, but they're a tremendous resource since they will do anything a TBC will do, plus lock up a nonsynchronous source.

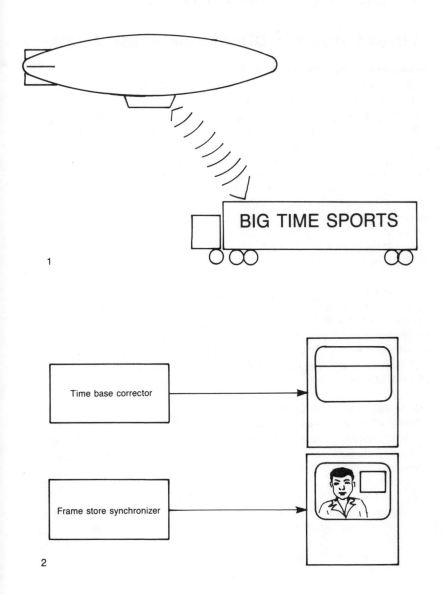

1. Since the blimp and the remote truck are operating on separate sync generators, they are nonsynchronous.

2. A TBC feeds out one line at a time while an FSS feeds out a field at a time.

117

Other Advantages of TBCs and FSSs

Dynamic tracking heads

TBCs and FSSs can give you a couple of other capabilities as well. If your VTR has a dynamic tracking head, a TBC will allow you to do broadcast-quality slow motion. A *dynamic tracking head* is a video head that adjusts itself to follow the video track when the tape changes speed. When you slow down the speed of videotape playback, the relationship between the video head and the angle of the video track changes. The dynamic tracking head automatically compensates for this change and always stays centered on the video track. A TBC or FSS simply maintains the proper sync with the rest of the system, even if, as in this case, the tape speed is slowed.

Freeze frames

A frame store synchronizer will also let you freeze frames. When you push the freeze on an FSS, it will continuously feed out the same field of video information. Some units will also feed out a complete frame, but then you often get frame jitter. This is because the video picture is not static. In the camera, after the electron gun finishes scanning a field, it returns to the top of the image to start the next field. But in the meantime, the subject of the picture may have moved just a little. So when the second field is interlaced with the first one, the subject may be offset a little and this can cause a jitter of the subject between one field and the other.

TBCs, VTRs, and production

TBCs are what have really made helical VTRs practical for broadcast use. TBCs and frame store synchronizes have become an integral part of TV production facilities everywhere, and the use of other digital equipment is becoming more and more commonplace.

Digital technology provides another advantage beyond what might be apparent from this discussion of TBCs and FSSs. Unlike analog, when the digital signal is run through several pieces of equipment, there is no increase in noise. Since digital information is a series of numbers, the D-to-A converter will ignore any surrounding noise when it converts that number to its corresponding voltage.

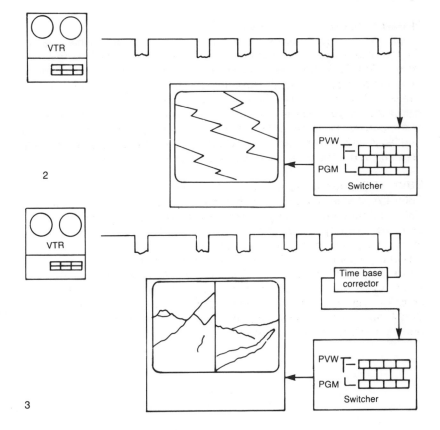

Video track laid down on tape

Path of video head

Normal speed or slow motion on a machine with a dynamic tracking head

Slow motion without dynamic tracking head

1

VTR

2

PVW
PGM

Switcher

VTR

Time base corrector

PVW
PGM

Switcher

3

1. **Dynamic tracking heads.**

2. **Integrating helical VTR playback without time base correction.**

3. **Integrating helical VTR playback with time base correction.**

Computer graphics allow you to create complex effects simply.

Computer Graphics for Video

Originating computer graphics

As you have already learned, digital video is really a series of numbers between 0 and 255 representing distinct voltages; the numbers are converted back to distinct voltages to form analog video. So, if you could sit down at a computer terminal and enter a series of about 250,000 numbers between 0 and 255, you would have just created a frame of video information. Needless to say, this would take a high degree of technical training and knowledge as well being a very time consuming and tedious way to do things. What is needed then, is a method of creating digital video images that would be as simple as typing at a typewriter or drawing with a pencil and paper.

Interface between people and machines

A number of digital graphic systems have developed that enable a person to take advantage of the opportunities offered by digital video. The capabilities of these systems range from producing simple letters and numbers to creating complex, detailed original illustrations and manipulating images on the screen. These systems provide an interface between the process of entering a series of numbers and the person who needs to get a job done. There are two things needed to make these interfaces successful. First, they must operate in a manner similar to what the user is already familiar with. If someone is introduced to a new piece of equipment that has several familiar features, he or she won't be as intimidated by it. But if a new piece of equipment is totally foreign, people will be more reluctant to try it out. Second, this new equipment needs to be flexible. The more that can be done with it, the more attractive it will be.

1. To effectively create a picture (or frame of video) on a computer, you might have had to enter thousands and thousands of values.

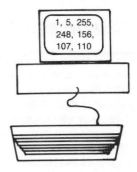

1

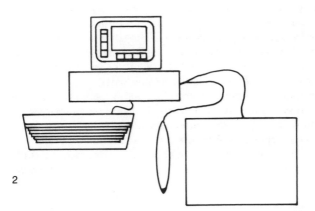

2

2. Fortunately, computers can translate our general commands into these thousands of specific values. We deliver these commands to the computer in simple ways, such as through a keyboard, an electronic "pencil and paper," or even voice.

121

How to create letters and numbers by computer.

Character Generators

Character generators (CG) were the first digital graphics units. A CG looks something like a computer terminal or a funny looking typewriter. It is, in fact, intended to be used like a typewriter. Many of the editing features now used on electronic word processing systems were first used on CGs. CGs allow you to type titles on the TV screen. If you want to make credits for who wrote, produced, directed, and starred in the show, use the CG. If you're interviewing someone and you want to flash his name up on the screen so the viewers know who he is, use the CG.

The earliest CGs were very simple affairs. They only produced white lettering on black background for use as keys. There were one or two type sizes, and only one or two type styles (*fonts*). But now you can do white on black lettering, or you can give the letters any color you want and produce a colored background. You can make the letters an outline, or solid, or solid with an outline. There's a wide variety of sizes and fonts available; the range is amazing.

To be most effective, character generators also need to have an extensive memory. CG memories are measured in pages. One complete video frame of information is a page. If you're preparing a show that has opening titles, names inserted during the show, and closing credits, you're going to need a lot of information. If the show is shot live, this is too much to be typed as it's needed. However, you can enter all the information into the CG's memory before the show begins. Then it can be brought out of the memory when needed. Many CGs will have an internal memory of several pages and an external computer-like disc memory that will hold several hundred pages. Of course, what is really happening is that the CG is generating a series of numbers between 0 and 255 that are going to a D-to-A converter that emits an analog video signal from the CG. But the main control unit will still look very much like an electric typewriter's keyboard or a standard computer terminal.

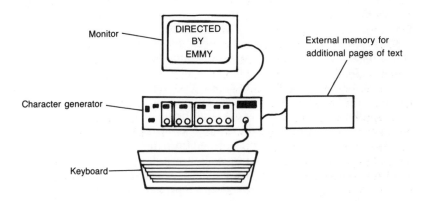

Monitor ⎯⎯⎯⎯⎯⎯⎯

DIRECTED
BY
EMMY

External memory for
additional pages of text

Character generator ⎯⎯⎯⎯⎯

Keyboard ⎯⎯⎯⎯

Components of a typical character generator.

Draw and paint on video.

Creating Imagery and Effects

Computer-generated imagery (CGI)

As the name implies, these digital graphics devices allow you to easily compose drawings or combinations of illustrations and text that can be manipulated and reproduced in a number of different forms, including on paper and as part of a video. These electronic graphic systems look very much like computers. There will be the terminal, with the monitor (maybe two) and the disc drives, but there will also be something that looks like a pencil and paper. It won't be a real pencil and paper, of course, but an *electronic palette* and *stylus*. Most of the work will be done with the palette and stylus so that an artist doesn't have to learn a lot about computers in order to use the system. The palette looks like a flat rectangular piece of plastic with a wire coming out of it. The stylus looks like a rather fat ball point pen with a wire coming out of the top of it. The monitor will display a rectangular image area and menu boxes (lists of commands by which the computer receives instructions). When the point of the pen is touched to the palette, a cross usually appears on the monitor. As the point of the pen is moved across the palette, a corresponding line appears in the image area of the screen. So, while you draw on the palette, the image doesn't appear there, but rather appears on the monitor. You can choose different brush styles and textures, colors, and other details by simply touching the point of the pen to the appropriate menu box on the monitor. The keyboard terminal is used to enter text into the art work, and the finished product can be stored and recalled from memory at will. Each make and model of electronic graphics system will have its own features, but most will operate along the lines outlined here. Like the character generator, the system is storing an image as a long stream of numbers that will be converted to analog voltages at the output of the system.

Digital video effects

Digital video effects (DVE) systems enable a user to manipulate a digitized video signal and produce any of an incredible array of special effects such as those described earlier. These devices can be incredibly complex, and the specific operation of different models can vary greatly. However, you should understand the basic concepts underlying their place in any video system. Remember that a DVE still serves as an interface between people and numbers. The video comes out of the switcher and into the DVE where an A-to-D converter changes it to digital information. This digitized video is then modified by the proper manipulation of knobs, levers, and buttons on the DVE control panel. From there it goes to a D-to-A converter and back to the switcher.

Digital video has opened vast new capabilities in TV production. Things that were unthought of five years ago are commonplace now. And any time the technology starts to become a little mind boggling, it might help to remember that it's just a computer throwing around a bunch of numbers.

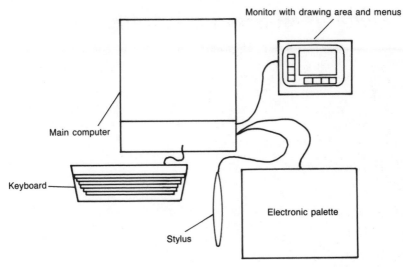

Monitor with drawing area and menus

Main computer

Keyboard

Stylus

Electronic palette

1

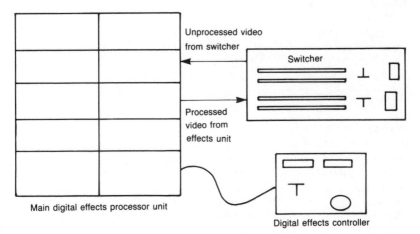

Unprocessed video from switcher

Switcher

Processed video from effects unit

Main digital effects processor unit

Digital effects controller

2

1. **Typical components of a computer graphics system.**

2. **Typical digital effects setup.**

Problems of Camera Pickup Tubes

When cameras and pickup tubes were discussed in an earlier section there wasn't really any discussion about some of the problems with pickup tubes. After all, if they're the only things you've got to work with, why worry about things you can't do anything about. But technology is developing a new method of changing light into video. In order to evaluate this new device we should take a closer look at some of the problems of pickup tubes.

Electronic problems

Pickup tubes are an old technology. The original designs go back almost 60 years. They're also relatively large and expensive. By the time you wrap the deflection coils around the tubes to direct the electron beam and add the control circuitry for those coils, you've added even more size, weight and expense. In addition, pickup tubes wear out. You might think that everything wears out, but much of today's solid state electronics can last almost forever.

Visual problems

Certain types of pickup tubes also create visual problems. Bright objects will literally burn off the light-sensitive coating on the target of some types of pickup tubes. This is called *burn in.* With some tubes, you'll get *streaking* in very high contrast situations. You've probably seen night football games on TV where the camera catches some of the stadium lights. As the camera moves from the lights back onto the field, you'll see red streaks, called comet tailing, follow the path of the camera. That's a characteristic of the pickup tubes. So, there are problems with pickup tubes that would be nice to eliminate.

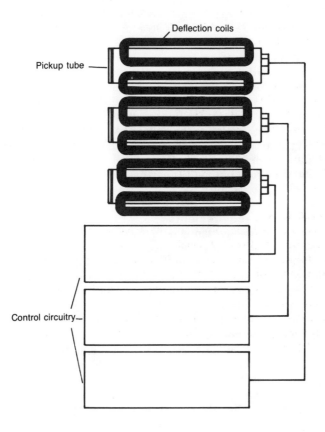

Deflection coils

Pickup tube —

Control circuitry—

Complexity of camera pickup tube system.

Charge Coupled Devices

There is a device that has been developed that solves most of the problems with pickup tubes. CCDs are developed from the same sort of solid state silicon chip technology that has made computers faster, smaller, more powerful, and cheaper; put powerful radios into small packages; and even made TBCs and DVEs possible for television.

CCD layout and operation
These chips are laid out with horizontal and vertical photo-sensitive elements (lines). Where the elements cross (called crosspoints) are our picture elements *(pixels)*. When light hits one of these pixels, a distinct electrical voltage is created. The brighter the light, the higher the voltage; the darker the light, the lower the voltage. All of these discrete voltages are read off left to right and top to bottom into a memory. The memory is then fed out horizontal line by horizontal line in sync with the rest of the system. Once the face of the CCD has been cleared, a new image forms and the process is repeated.

Broadcast-quality requirements
To meet broadcast-quality standards, a CCD needs to have around 250,000 pixels. Since having more pixels gives higher detail and resolution, it would be nice to have even more than that. Cramming that many pixels into an area a little larger than your thumbnail has been one of the big problems. Engineers have been working on this problem for about ten years and they've finally been successful.

CCD production problems
Although high-density CCDs have been developed, getting them produced in economical quantities has been a problem. If only a small percentage of the chips produced are usable, then the cost per chip is pretty high. And that's been one of the problems. Very few chips out of a production run were meeting the standards, therefore most of the batch had to be discarded. Many of the chips that were discarded before can now be used in less expensive or consumer cameras. It also looks like things are being reversed and many more of the chips being produced can be used; thus the cost per chip is getting cheaper.

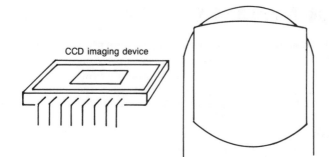

1. CCD next to a thumb; the scale is roughly accurate.

2. CCD layout. This diagram shows horizontal and vertical layout of a CCD. Every time a horizontal line crosses a vertical line, a pixel is created.

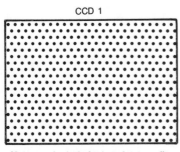

CCD 1

Not enough pixels for broadcast quality

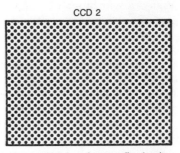

CCD 2

Approaching broadcast quality density

3. Broadcast quality CCDs need a high pixel density.

4. In the past, CCDs were expensive because each batch produced only a small number of usable chips.

Usable chips

Low % of usable chips

High % of usable chips

CCD Advantages

Now that these problems of high density and production have been solved, the CCD has some real advantages over traditional pickup tubes. As mentioned earlier, they're not much bigger than your thumbnail, they don't have any electron beams (so there is no need for deflection coils to direct an electron beam), and there's no need for circuitry to control the coils. This takes quite a bit of size and weight away from a typical tube-type TV camera. Broadcast-quality cameras and recorders might be smaller than a shoe box and weigh less than five pounds.

CCD field cameras
There are a number of broadcast-quality CCD field cameras available, which produce very good pictures. Indeed, some engineers suggest that within a couple of years tube-type cameras will no longer be made.

Economic advantages of CCDs
CCD chips should cost much less than a top-quality pickup tube does. Broadcast-quality cameras should become much smaller, lighter, and less expensive. And to add to this advantage, CCDs should last almost forever. They won't need to be replaced on a regular basis like pickup tubes.

Visual advantages of CCDs
If you're still not convinced, CCDs don't have most of the visual problems of tubes such as burn-in and streaking.

Most engineers agree that it's not a question of whether CCDs will replace traditional pickup tubes, but when. There's little doubt that once the density and production problems are completely solved that the television pickup tube will go the way of black and white TV. And when was the last time a TV show was produced in black and white?

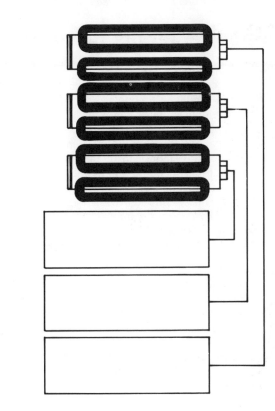

CCDs

1

Pickup Tubes	CCDs
Relatively large	Small
Expensive	Relatively inexpensive
Require deflection coils and control circuitry	No deflection coils or control circuitry required
Wear out	Don't wear out
Low contrast ratio	High contrast ratio
Burn in	Don't burn
Excellent picture quality	Good picture quality (and improving!)

2

1. Pickup tubes, deflection coils, and related control circuitry.

2. Comparison of pickup tube characteristics to CCDs.

A studio usually has one of several video signal systems.

Composite Versus Component Video

Problems of composite video

In a previous section you learned that encoding is the process of combining the three chrominance channels and the luminance channel into a single composite video signal. Many pieces of video equipment need to decode this composite signal into its component parts in order to process and use it. Of course, the component signals must then be re-encoded before the composite signal can be sent on its way from that piece of equipment. All of this decoding and re-encoding creates problems. Every time the signal is encoded or decoded it is distorted a little and a little noise is added. The system has worked well for the last 40 years or so, but now as creative people want to use more sophisticated production techniques and consumers want a better picture, demands are being made for better-quality video. There are a number of approaches to the problem.

Component video

In the section on switchers you learned about component switchers that have separate red, green, and blue (RGB) inputs for each channel. Taking the separate RGB outputs from each camera and running them directly to the switcher bypasses the entire encoding circuitry of the camera. This provides a cleaner, sharper picture. Keys come across especially well with this system. Of course, to get the most out of this system the entire studio needs to be component-oriented — cameras, distribution amplifiers, switchers, and character generators. This will probably require re-engineering the entire control room since three cables will now have to be installed for every one that was there before. Timing the system becomes even more crucial.

Most equipment that originates video, such as cameras and character generators, is easily integrated into a component system. This equipment already generates RGB signals, so it doesn't take much to bypass the encoder. The DAs and switchers, however, are much more expensive. As a result, it is going to take a big budget to convert the entire studio to component video.

Y/C

A less expensive component system that produces higher quality video than composite video is *Y/C*. In technical terms, Y stands for the luminance or black and white information and C stands for the chrominance or color information. Using a Y/C system eliminates much of the encoding and decoding problems inherent in composite video.

To get an idea of how this system works, take a look at Y/C VTRs. Rather than lay down an entire field of composite video in one track, a Y/C must lay down the Y and C information separately to make a field. This provides much better quality at a reasonable cost. Because of these advantages, many feel that Y/C will completely replace composite video in the near future.

132

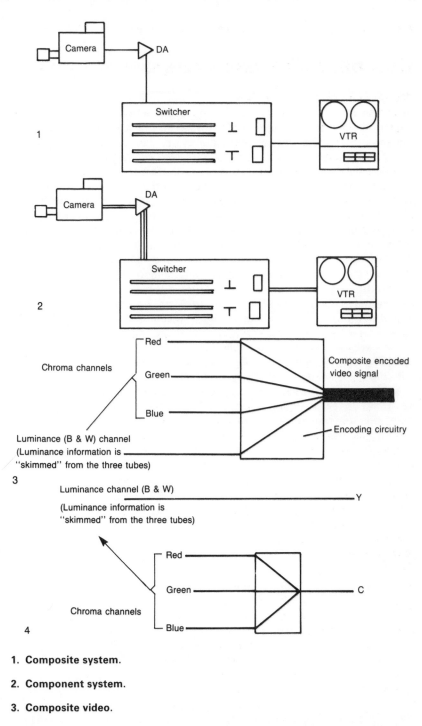

1. **Composite system**.

2. **Component system**.

3. **Composite video**.

4. **Y/C video**.

Complete Digital Systems

What might seem to be the obvious solution to everything would be to make the whole system digital. After all, if you immediately convert the analog video signal to digital right in the camera, all that is happening thereafter is that you are sending a bunch of numbers through the cable for computers to process. Even if you have to convert that digital information back to an analog signal at the broadcast transmitter you're still going to have a vastly superior signal. With a totally digital system, there's not going to be the constant encoding and decoding going on and, in the system described above, the D-to-A and A-to-D conversations are going to take place only at the beginning and end of the system.

The problem with this solution is that in the past it has not been practical. There are literally hundreds of makes and models of digital equipment out there with little uniformity among them. Different manufacturers used different sampling rates,different methods of processing, and different approaches to solving the same problems. As a result, one piece of digital equipment would not necessarily directly hook up to another. With all the time and money spent in development and marketing, it wasn't likely that manufacturers were going to throw their current product lines out the window in order to develop new products that would interface with the equipment of other manufacturers.

Now, however, manufacturers are building equipment that meets internationally accepted digital standards. As more of this compatible equipment hits the marketplace, you'll start to see fully digital studios that produce incredibly high quality pictures.

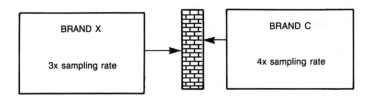

Incompatibility between different makes of equipment.

Digital Videotape Recorders

There is great interest in digital videotape recorders because of their ability to be duplicated through many many generations without signal degradation or increased noise. There are currently two digital tape formats available that have found different uses in the market.

D-1 component

The *D-1* system is a component digital system. This machine won't even accept composite video. Only a digital signal can be fed into it. If you take the advantages of the analog component system discussed above and combine them with the advantages of the digital system discussed, you have amazing potential. This is exactly the type of system the D-1 digital videotape recorder was designed for. Since you have to run three channels of everything, this system is extremely expensive. It does, however, produce incredible pictures. The D-1 machine is particularly useful for very sophisticated production techniques where one frame of video at a time is laid down on the tape or where multilayered keys and special effects are needed. Professional video production houses, where top-quality video is an absolute necessity, are the most likely users of this machine. It's highly unlikely that very many local stations will make much use of the D-1 format.

D-2 composite

The *D-2* composite system takes the single composite analog video signal and converts it to digital, or it can take a direct composite digital input. This digital signal is then laid down on the tape. The output of the D-2 machine is either composite analog or digital video. This allows the D-2 to be easily integrated with standard studio equipment. Several D-2 VTRs could also be hooked together for direct digital-to-digital editing. This is a much simpler and far less expensive system than D-1, yet D-2 is far superior to traditional analog video recording technologies. Many people feel that D-2 will become the standard videotape format of local TV stations.

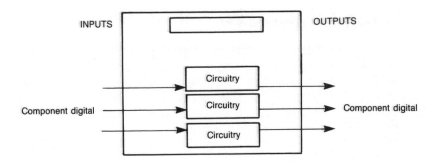

1

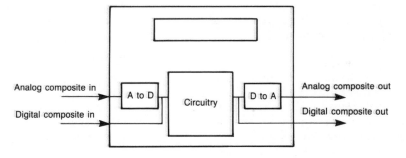

2

1. **D-1 component digital VTR.**

2. **D-2 digital tape.**

High-Definition TV

As good as these digital systems are, even the best leave something to be desired. You may have seen very large screen TVs and noticed that their pictures weren't as sharp and clear as smaller TV sets. That's because you have got the same 525 lines spread out over a much larger area. Anytime you spread a given amount of information over a large area, it's not going to look as sharp and clear, even in digital formats. This really demonstrates how much quality is lacking in a TV picture when compared with film. For many years there has been a great deal of research and development in TV systems that would produce a picture that approaches film quality. This is called high definition TV (HDTV). At the time of this writing, its hard to tell what's going to happen in the field of HDTV because there are more than a dozen different formats out there. A very good system that uses 1,125 lines and 30 frames and has a wider picture has been available for some time, but only a few production houses have completely converted to it. This format may become the standard production format. The problem is that it is incompatible with home TV. It would take a standard broadcast station two regular TV channels to broadcast the signal and a normal (by today's standards) TV set couldn't pick it up. You could pick it up with a special TV set, but that set couldn't pick up regular TV signals. Several systems have been proposed that would be compatible with regular TV, but only time will tell what ultimately will happen in this field.

1. Home TV now.

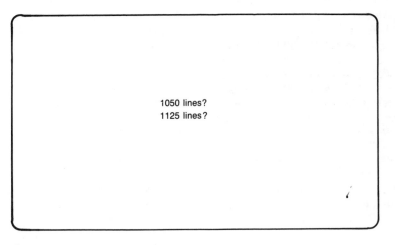

525 lines

1

1050 lines?
1125 lines?

2

2. HDTV will have a wider screen and more scan lines.

Patch Panels

Patch panels are an extremely basic item of equipment that are not always easy to understand. Patch panels help interconnect the video system components. For this reason, to successfully use them you must understand all of the other components and be able to carefully reason and evaluate possible connections and their results.

Because patch panels are such a vital piece of equipment, this special section both explains their use and offers some basic exercises. Remember that, although you will grasp the fundamentals by reading this text, to become fully comfortable with patch panels you should practice on your own with real equipment.

What patch panels do

Patch panels are routing devices that allow you to take the signal from one place to another. They don't change, convert, or do anything else to the signal. The panel is made up of a group of connectors (*jacks*), and if you want to hook two pieces of equipment up, you simply plug a cable into the output of one piece of equipment and into the input of another.

Patch panel components

Here's a quick example. There'll be three pieces of equipment you'll be dealing with: (1) The panel itself with holes (jacks) in it (each jack will be labeled); (2) the *patch cable,* which is made up of a video cable with a connector (*plug*) at each end designed to fit into the jacks; and (3) a *hairpin,* little more than a U-shaped patch cable encased in a small block of plastic. It is used to connect jacks adjacent to each other on the patch panel. The first figure is a very small patch panel. It only has jacks for six inputs and outputs. You'll notice that this patch panel has all the outputs along the top and all the inputs along the bottom, but not all patch panels are set up this way.

Patching example

Assume that you want to record the output of camera 2 on both VTRs 1 and 2. You would hairpin camera 2 into DA 1, then you would take the first DA output and hook it into the input of VTR 1 and the second output of DA 1 and hook it into the input of VTR 2. The panel would appear something like the last figure.

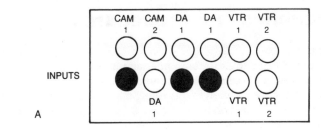

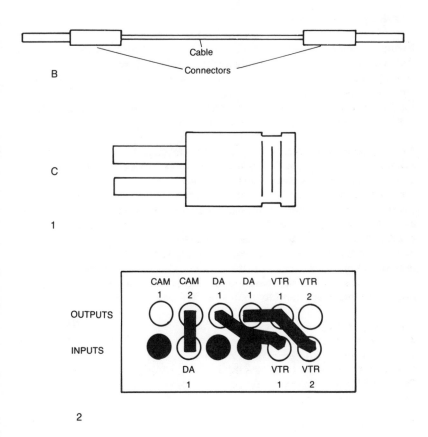

1. Patch panel components. (A) Simple patch panel; (B) patch cable;
(C) hairpin.

2. Diagram of the patching example.

141

Patching Rules and Procedures

Patching rules

There are a couple of basic rules to remember about using patch panels.

1. Most pieces of equipment have only one output and one input. The DAs are the most notable exception to this rule. So if you need to take one signal to more than one place, you have to use a DA as we did in the above example.

2. Outputs are always hooked up to inputs. Never output to output or input to input.

Patching procedures

There is also a procedure to follow that may have up to five steps when trying to decide how to make a patch. The procedure itself is simple, but the questions it raises are not always that easy to answer.

1. What is your starting point? What piece of equipment is your signal coming from?

2. What is your destination? In other words, where do you want to go and where do you want the signal to end up?

3. Is there anything in between? Does any equipment need to go between the starting point and destination pieces of equipment? If the answer is no, then just patch between the pieces of equipment in steps 1 and 2 above. If the answer is yes, go to step 4.

4. What piece or pieces of equipment go between the starting point and the destination? If there is only one piece of equipment, you take the output from step 1 above to the input of the answer to this question. The output of that is taken to the input of the piece of equipment in step 2 above. If it takes more than one piece of equipment to answer this question, go on to step 5.

5. In what order? You need to decide what order the intermediate pieces of equipment need to go in since the patch must be sequential.

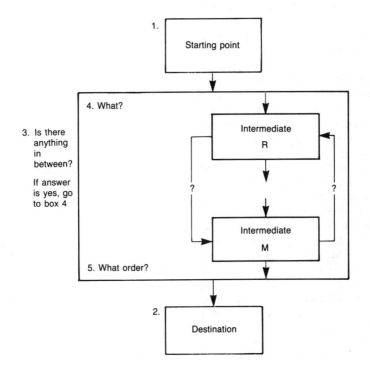

1. Starting point

4. What?

3. Is there anything in between?

If answer is yes, go to box 4

Intermediate R

?

?

Intermediate M

5. What order?

2. Destination

Steps for solving Patch Panel Problems.

Explanation of a Small Patch Panel

Here's an example of a small patch panel to see if you can figure out some patches using the procedure above. As with the previous patch panel, this one has all the outputs along the top row and all the inputs along the bottom (first figure). Go through all the jacks to make sure you understand the labels. Jacks 1 and 2 are the outputs of cameras 1 and 2. Output 3 is from the film chain. Output 4 is from the character generator. Output 5 is not being used. Outputs 6, 7, and 8 are from the program DA. In other words, they represent three identical outputs of the switcher. Output 9 comes from VTR 1, and 10 is the TBC output. Output 11 is not being used and 12 comes from VTR 2.

The lower jacks are the inputs. Jacks 13 and 14 are not being used. Jack 15 is the input to the CG. Jack 16 is the input to the production switcher that is labeled CG. Input 17 goes to the engineering switcher (this is a passive switcher that allows the engineers to punch up whatever they want to look at in the system without interfering with the production). Jack 18 is hooked up to a video connector in the studio in case a signal needs to be fed to some piece of equipment on the studio floor. Jack 19 is the input to the studio monitor so that the program can be seen in the studio. Input 20 goes to VTR 1 and input 21 goes into the TBC. Input 22 goes into the production switcher button labeled VTR 1, and 23 is the input to VTR 2. Jack 24 is not being used.

Normal setup

There are a number of these jacks that you will want to have hooked up in normal use (second figure). For example, you probably want the CG to go to the switcher so that you can use titles during a production. So you would use a hairpin to connect jack 4 to jack 16. In addition, the program output (jack 8) is patched to jack 23 so that the program can be recorded on tape. You also want to be able to play back a tape through the switcher, but need to time base correct the signal before you can do that, so jack 9 is patched to jack 21. In order to get the time base–corrected VTR signal into the switcher, you patch one end of the cable into jack 10 and the other end into jack 22.

Look at this patch panel for a minute and see if there is anything else you might want to hook up. It would be nice if the talent in the studio could see what was being recorded. What patch would you make in order to get that result? Go through the procedure again. What are your starting point and destination? Since the program DA output is what is being recorded, that's what you want the talent to see, so that's your starting point. You already have a monitor in the studio, so that's your destination. Is there anything that goes between? The answer is no, so you can just patch the program output to the studio monitor. Since jack 8 is already being used you can use either jack 6 or 7. Jack 7 has been hairpinned to jack 19 in the last figure. This would give you a setup of the patch panel that could be used in most normal shooting situations.

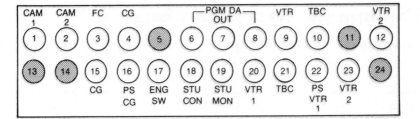

2

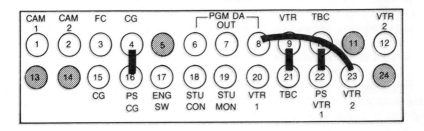

3

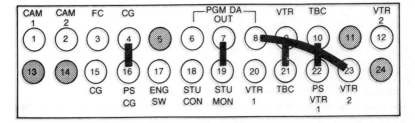

1

Patch panels will seem easier after you work with them for a while.

Simple Patching Exercises

Now, referring again to the basic setup on page 144, assume you have a different situation. The director wants to record the output of each camera on a separate VTR. That will allow him or her to select the best shots from each camera and edit them together at a later time. This really presents two problems since you have two starting points (cameras 1 and 2) and two destinations (VTRs 1 and 2).

It would be easier if you could pull all the cables and hairpins out of the patch panel and start with a clean panel. However, that is rarely done. Most patch panels are left in their normal configurations. Any new patches must be made from that configuration. Besides, if you pull all of the cable out of the patch panel, you might disconnect equipment that is being used elsewhere in the studio. Because of this, the problems presented here will assume that the patch panel is in its normal configuration.

In the problem presented above nothing needs to go between the cameras and VTRs, so you can hook them up directly. But first, you have to unhook the program output from VTR 2 since only one cable can be patched into a jack at a time (first figure).

This time you must solve a harder problem. Assume that your switcher had broken down. As a result, there's nothing coming out of any of the program DA outputs. But you want to edit between VTRs 1 and 2 and add titles in the process. Go through the procedure again. What's your starting point? How about VTR 1? Now what's your destination? VTR 2 will do just fine. Is there anything between? Yes, because you want to add titles. What? Since you need titles, the CG probably goes between. And it probably wouldn't hurt to stabilize the playback signal so you probably want the TBC as well. In what order? Do you want to stabilize the video before it goes into or after it comes out of the CG? Unstable video certainly won't help the CG, so it's probably a good idea to stabilize it first. That leaves you with the flow diagram in the second figure.

Since the hairpin is already between the VTR 1 output and TBC input, you don't have to patch that. But you do need to pull out the hairpin between the CG output and the production switcher, and you also need to remove the cable between the program output and VTR 2 input. Then all you have to do is take the TBC output to the CG input, and the CG output to the input of VTR 2 (last figure).

Summary of patch panels

The patch panel that you've practiced on is intentionally small and simple so that you can develop a basic understanding of the equipment. Although patch panels in a typical TV station are much larger and complex, the underlying principles are the same. Take your time when working on a patch panel and think the problem through carefully.

Patch panels tie the video system together. When you understand patch panels and are comfortable with them, you will truly understand the entire system.

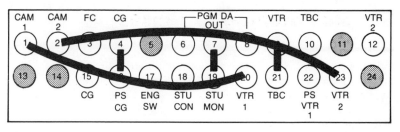

1

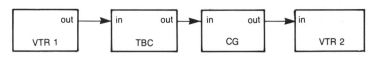

2

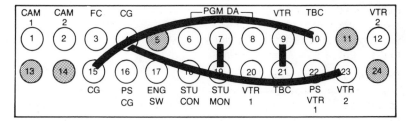

3

Further Reading

ALTEN, STANLEY:
Audio in Media, 2nd ed. Belmont, CA: Wadsworth, 1986.

ANDERSON, GARY H.:
Video Editing and Post-production: A Professional Guide, 2nd ed. White Plains, NY: Knowledge Industry Publications, 1988.

BROWNE, STEVEN E.:
Videotape Editing: A Postproduction Primer. Boston and London: Focal Press, 1989.

MILLERSON, GERALD:
Effective TV Production, 2nd ed. Boston and London: Focal Press, 1983.

MILLERSON, GERALD:
The Technique of Lighting for Television and Motion Pictures, 2nd ed. Boston and London: Focal Press, 1982.

MILLERSON, GERALD:
The Technique of Television Production, 12th ed. Boston and London: Focal Press, 1990.

ORINGEL, ROBERT S.:
Television Operations Handbook. Boston and London: Focal Press, 1984.

SCHNEIDER, ARTHUR:
Electronic Postproduction and Videotape Editing. Boston and London: Focal Press, 1989.

WATKINSON, JOHN:
The Art of Digital Audio. Boston and London: Focal Press, 1988.

WURTZEL, ALAN and ACKER, STEPHEN R.:
Television Production, 3rd ed. New York: McGraw-Hill, 1989.

ZETTL, HERBERT:
Television Production Handbook, 4th ed. Belmont, CA: Wadsworth, 1984.

Glossary

Additive colors The color system that mixes colored light to create all the various colors of the color spectrum.

Amperes (amps) The unit of measurement for current.

Analog video signal The varying voltages that make up the video information of a television signal.

Assemble edits Edits that lay down all aspects of the signal: audio, video, and control track all at the same time.

A-to-D converter The circuitry that converts analog signal information into digital information.

Audio-follows-video switcher A switcher that changes both audio and video sources with the push of one button.

Back porch The portion of the waveform scan that represents the horizontal blanking just before the start of a new line of video.

Black burst A signal from the sync generator that includes all normal blanking and sync information along with black video.

Blanking That time when the electron guns in the system are turned down to a low voltage so that they can return to the beginning of a new line or field.

Blanking pulses Signals from the sync generator that signal the camera's electron gun to go into blanking.

Bus A row of buttons on a switcher that allows a person to change between various video sources that are available in the system.

Camera A device that changes light images into a usable electronic signal.

Capstan servo See **Vertical Lock**

CCD (Charge Coupled Device) A solid-state device used, instead of a pickup tube, for changing light images into an electronic video signal.

Character generator (CG) A machine that creates words and titles for the TV screen.

Chroma The color information in a TV signal.

Chroma key A special effect where a chosen color is replaced with video from another source.

Chroma key tracking A digital effect that compresses the signal from a video source into the available chroma key window.

Color burst or **3.58** The color reference that is inserted in every video line that determines how color information is to be interpreted.

Color subcarrier See **Color burst**

Color sync See **Color burst**

Complementary colors The colors cyan, magenta, and yellow that are created by mixing parts of the primary colors (red, green, and blue).

Component switcher A video switcher that deals with the individual color components (red, green, and blue) of the picture instead of the encoded composite video signal.

Component video A video signal made up of the individual component parts as opposed to an encoded composite signal.

Composite video The video signal made up of both the video and sync information.

Compressions Digital effects where the size and/or aspect of the picture is changed on the TV screen.

Control track A track on the videotape used to help stabilize tape playback speed.

Control track counter editing controller A device that controls videotape editing by counting the control track pulses on the tapes.

CRT (Cathode ray tube) Television picture tube.

Current The volume of electrons passing a given point at a given time, measured in amps.

D-1 A system of digital videotape recording that uses component video.

D-2 A system of digital videotape recording that uses composite video.

DA (Distribution amplifier) A piece of equipment that produces multiple outputs identical to its input signal.

D-to-A converter Circuitry that converts digital information into analog information.

Deflection coils Electromagnetic coils around a camera pickup tube used to steer the tube's electron beam in the scanning of its target.

Deflection yoke Electromagnetic coils around a CRT used to steer the CRT's electron beam as it sprays the picture across the screen.

Digital video A video signal that is made up of a series of assigned numbers rather than analog voltages.

Drive pulses Signals from the sync generator that control the scanning of the electron beams.

Dubbing The process of copying the electronic signal from one tape to another.

Dynamic tracking head A videotape head that automatically aligns itself with the center of the video track on the tape for slow motion or freeze frames.

Encoding The process of combining the chroma and luminance information into a single signal.

Fields The complete set of odd- or even-numbered lines that, when interlaced together, make up one video frame.

Film chain The system, using a standard video camera, that converts film (motion pictures and slides) into a video signal.

Flow diagrams A diagram that uses geometric shapes and lines in place of equipment and wires to illustrate the interconnection of equipment.

Flying spot scanner   A film-to-video transfer device that uses the electron beam of a CRT to illuminate the film being transferred.

Frame   In the American system, two interlaced fields of 262.5 lines each that, when combined, make a complete picture of 525 lines. There are 30 frames per second in the NTSC (American) system.

Frame lock   A method of stabilizing videotape playback that tries to match an even field to an even field and an odd field to an odd field of the playback signal to that of the signal coming from the sync generator.

Frame store synchronizer (FSS)   A device used to lock up nonsynchronous video signals to the main system.

Front porch   The portion of the waveform scan that represents the horizontal blanking at the end of a line of video.

Giga   (G)   The abbreviation for billions (1,000,000,000). For example, 6 GHz would equal 6,000,000,000 Hz.

Gyroscopic time base error   Time base error that is created when a videotape recorder is moved perpendicular to the plane of the head drum's rotation.

HDTV   High-definition TV.

Head   The small electromagnetic device that lays down or picks up the information on a piece of recording tape.

Headwheel   The rotating disc on which the video heads are mounted.

Helical   A method of video recording that lays down video information at a slant to the tapes direction of travel. Also known as slant track recording.

Hertz   (Hz)   A measurement of frequency equal to one cycle per second.

Horizontal blanking   The time when the electron guns are turned down to a low voltage at the end of a line until they are turned back up at the beginning of a new line.

Horizontal lock   A method of stabilizing videotape playback that tries to match a horizontal sync pulse of the playback signal to each horizontal sync pulse coming from the sync generator.

Horizontal sync   The signal from the sync generator that causes the electron gun to return to the other side of the screen for a new line.

Impedance   A measurement of the properties that tell whether two or more circuits will interact well. Measured in ohms.

Induction   The process whereby a circuit with a stronger magnetic field forces some of its signal into a circuit with a weaker magnetic field.

Insert edits   Edits that use control tracks that have already been laid down on the tape to be edited to.

Interlace scanning   The process of taking a field of odd-numbered lines (262.5) and combining it with a field of even numbered lines (262.5) to make a complete video frame (525).

Keys A special effect whereby the signal from one video source "cuts" a hole into another video source.
Kilo (K) The abbreviation for thousand (5 K = 5,000).

Luminance The black and white portion of the video signal.
Luminance keys A key whereby the hole being cut is determined by brightness of the video source.

Manual editing Editing that is completely done by a person without using an electronic editing controller.
Matte key A luminance key where the "hole" created by the key is filled with artificially created color from the switcher.
Mega (M) The abbreviation for million (3 M = 3,000,000).
micro (μ) The abbreviation for millionth (5 μ = 5/1,000,000).
milli (m) The abbreviation for thousandth (200 m = 200/1,000).
Mismatch Refers to impedance where two pieces of equipment will not work well together.
Monitor A TV set designed to display a straight video signal as opposed to a set designed to receive programs off the air.
Multiplexer A system of mirrors on a film chain that allows the projection of one of several film sources into the TV camera.

nano (n) The abbreviation for billionth (5 n = 5/1,000,000,000).
Noise Unwanted electromagnetic static inherent in all electronic circuits.
Noncomposite video Video information without sync information.
Nonsynchronous A signal that is completely out of sync with the main system.

Off-line editing The process of developing an editing script made up of SMPTE code numbers and a work print on small-format helical equipment.
Ohm (Ω) The unit of measurement for both resistance and impedance.
On-air switcher The switcher used to determine what goes to the transmitter. Usually an audio-follows-video switcher.
On-line editing The process of editing the final, finished tape on large-format tape machines.
Out of phase When cameras show different colors during a transition because their color bursts are not matched.
Oxides The coating on tape that allows signals to be recorded magnetically.

Passive switcher A switcher made up of several simple mechanical switches.
Patch panel A device that allows flexible routing of signals from one place to another.
Pedestal The black portions or areas of the TV picture.
Pickup tube A device that changes light images into an electrical signal.

Pixel Picture element.

Positive interlace An interlace method where the position of each line will be the same in every frame of video.

Primary colors In television, the colors red, green, and blue.

Proc amp Video processing amplifier, a piece of equipment that strips the distorted sync off of a videotape playback signal and replaces it with clean sync. Most proc amps also allow the control of some of the video parameters, such as hue, video brightness, pedestal (black) levels, etc.

Quantizing The process of converting a sample of video information into a number.

Random interlace A system of interlace scanning whereby the position of video lines scanned varies a little with each frame.

Registration The process of ensuring that the signals coming from the individual red, green, and blue pickup tubes are aligned properly.

Resistance A measurement in Ohms indicating the constraint to the flow of electricity.

Routing switcher A switcher used to route signals from one place to another.

Sampling The process of "grabbing" a piece of analog video information so that it can be quantized, or converted into numbers for either processing or storage in memory.

Self-fill key A luminance key whereby the hole cut is filled by the video that cut the hole.

Signal-to-noise ratio A comparison of the strength of the signal coming out of a piece of equipment in relation to the internal noise created by that equipment.

SMPTE Society of Motion Picture and Television Engineers.

SMPTE time code A digital code laid down on tape that gives each frame of video a unique and unchanging address (03:38:52:04 would equal 3 hours, 38 minutes, 52 seconds, and 4 frames).

Switchers A device that allows a person to make a transition between video sources.

Sync A signal from the sync generator that causes the electron guns to return to the beginning of a new video line or field.

Sync generator A device that provides various sync signals (drive pulses, blanking pulses, and sync pulses) that keeps all of the equipment in a video system working together.

Tape transport system The mechanical system that pulls tape through a machine at an even speed.

Target The photosensitive coating of a pickup tube.

TBC (Time base corrector) A device for correcting time base error in videotape playback.

Time base error   The instability of a videotape playback signal created by the machine's inability to play back at exactly the same speed at which the tape was recorded.

Timing the system   Ensuring that all of the sync pulses from the various pieces of equipment arrive at the switcher at the same time.

Vectorscope   A piece of equipment that shows a graphic display of the color portion of the video signal.

Vertical blanking   The time when the electron beams are turned down at the end of a video field until they are turned back up at the start of a new field.

Vertical interval   That period of time when the electron beams are in vertical blanking.

Vertical interval switcher   A switcher that delays cuts between video sources until the entire system is in vertical blanking.

Vertical lock   A method of stabilizing videotape playback that tries to match the control track pulses of the playback signal to vertical sync pulses coming from the sync generator. Also called capstan servo.

Vertical sync   The signal from the sync generator that tells the electron beams to return to the top of the screen for the start of a new video field.

Volt   The measurement for the pressure of electricity.

Voltage   The pressure of electricity, measured in volts.

VTR   (Videotape recorder)   A machine that records sound and pictures onto magnetic tape.

Watt   A measurement of electrical power.

Waveform monitor   A piece of equipment that shows a graphic display of the black and white portion of the video signal.

Window of correction   The amount of time base error a TBC will correct, measured in video lines.

Y/C   Y equals the luminance portion and C equals the chrominance portions of the video signal. A Y/C piece of equipment or system will keep the components separate as much as possible.